T0073114

A Guide to Graph Algorithms

Ton Kloks • Mingyu Xiao

A Guide to Graph Algorithms

 Springer

Ton Kloks
Computer Science and Engineering
University of Electronic Science
and Technology of China
Chengdu, Sichuan, China

Mingyu Xiao
Computer Science and Engineering
University of Electronic Science
and Technology of China
Chengdu, Sichuan, China

ISBN 978-981-16-6349-9   ISBN 978-981-16-6350-5 (eBook)
https://doi.org/10.1007/978-981-16-6350-5

This Springer imprint is published by the registered company Springer Nature Singapore Pte Ltd.
The registered company address is: 152 Beach Road, #21-01/04 Gateway East, Singapore
189721, Singapore

Contents

Preface   **XI**

Acknowledgments   **XV**

Graphs   **1**

   1.1   Isomorphic Graphs   2
   1.2   Representing graphs   2
   1.3   Neighborhoods   3
   1.4   Connectedness   4
   1.5   Induced Subgraphs   4
   1.6   Paths and Cycles   5
   1.7   Complements   7
   1.8   Components   7
          1.8.1   Rem's Algorithm   7
   1.9   Separators   10
   1.10 Trees   10
   1.11 Bipartite Graphs   12
   1.12 Linegraphs   14
   1.13 Cliques and Independent Sets   14
   1.14 On Notations   15

Algorithms   **17**

   2.1   Finding and counting small induced subgraphs . . .   18
   2.2   Bottleneck domination   20
   2.3   The Bron & Kerbosch Algorithm   22
          2.3.1   A Timebound for the B & K −Algorithm .   28
   2.4   Total Order!   30
          2.4.1   Hypergraphs   34

2.4.2   Problem Reductions   35

2.5   NP – Completeness   38

2.5.1   Equivalence covers of splitgraphs   39

2.6   Lovász Local Lemma   42

2.6.1   Bounds on dominating sets   46

2.6.2   The Moser & Tardos algorithm   48

2.6.3   Logs and witness trees   49

2.6.4   A Galton - Watson branching process   52

2.7   Szemerédi's Regularity Lemma   54

2.7.1   Construction of regular partitions   60

2.8   Edge - thickness and stickiness   72

2.9   Clique Separators   74

2.9.1   Feasible Partitions   76

2.9.2   Intermezzo   79

2.9.3   Another Intermezzo: Trivially perfect graphs   80

2.10   Vertex ranking   81

2.10.1   Permutation graphs   81

2.10.2   Separators in permutation graphs   82

2.10.3   Vertex ranking of permutation graphs   83

2.11   Cographs   84

2.11.1   Switching cographs   85

2.12   Parameterized Algorithms   88

2.13   The bounded search technique   90

2.13.1   Vertex cover   90

2.13.2   Edge dominating set   91

2.13.3   Feedback vertex set   92

2.13.4   Further reading   95

2.14   Matchings   96

2.15   Independent Set in Claw - Free Graphs   97

2.15.1   The Blossom Algorithm   97

2.15.2   Minty's Algorithm   100

2.15.3   A Cute Lemma   101

2.15.4   Edmonds' Graph   102

2.16   Dominoes   104

2.17   Triangle partition of planar graphs   106

2.17.1   Intermezzo: PQ - trees   110

2.18   Games   111

2.18.1   Snake   112

2.18.2 Grundy values 113

2.18.3 De Bruijn's game 114

2.18.4 Poset games 115

2.18.5 Coin - turning games 116

2.18.6 NIM - multiplication 118

2.18.7 P_3 - Games 120

2.18.8 Chomp 122

Problem Formulations **125**

3.1 Graph Algebras 125

3.2 Monadic Second – Order Logic 126

3.2.1 Sentences and Expressions 126

3.2.2 Quantification over Subsets of Edges . . . 127

Recent Trends **129**

4.1 Triangulations 129

4.1.1 Chordal Graphs 129

4.1.2 Clique – Trees 132

4.2 Treewidth 134

4.2.1 Treewidth and brambles 135

4.2.2 Tree - decompositions 137

4.2.3 Example: Steiner tree 139

4.2.4 Treewidth of Circle Graphs 145

4.3 On the treewidth of planar graphs 149

4.3.1 Antipodalities 151

4.3.2 Tilts and slopes 157

4.3.3 Bond carvings 163

4.3.4 Carvings and antipodalities 168

4.4 Tree - degrees of graphs 174

4.4.1 Intermezzo: Interval graphs 175

4.5 Modular decomposition 177

4.5.1 Modular decomposition tree 179

4.5.2 A linear - time modular decomposition 180

4.5.3 Exercise 187

4.6 Rankwidth 187

4.6.1 Distance hereditary - graphs 188

4.6.2 Intermezzo: Perfect graphs 190

4.6.3 χ - Boundedness 191

4.6.4   Governed decompositions 196

4.6.5   Forward Ramsey splits 198

4.6.6   Factorization of trees 199

4.6.7   Kruskalian decompositions 203

4.6.8   Exercise 204

4.7   Clustered coloring 205

4.7.1   Bandwidth and BFS - trees with few leaves . 205

4.7.2   Connected partitions 207

4.7.3   A decomposition of K_t minor free graphs . . 210

4.7.4   Further reading 211

4.8   Well - Quasi Orders 213

4.8.1   Higman's Lemma 213

4.8.2   Kruskal's Theorem 215

4.8.3   Gap embeddings 216

4.9   Threshold graphs and threshold - width 217

4.9.1   Threshold - width 218

4.9.2   On the complexity of threshold - width . . . 221

4.9.3   A fixed - parameter algorithm for threshold -
width 222

4.10 Black and white - coloring 227

4.10.1   The complexity of black and white - coloring   228

4.11 k – Cographs 229

4.11.1   Recognition of k – Cographs 231

4.11.2   Recognition of k – Cographs — revisited . . 232

4.11.3   Treewidth of Cographs 233

4.12 Minors 234

4.12.1   The Graph Minor Theorem 235

4.13 General Partition Graphs 236

4.14 Tournaments 240

4.14.1   Tournament games 240

4.14.2   Trees in tournaments 242

4.14.3   Immersions in tournaments 246

4.14.4   Domination in tournaments 255

4.15 Immersions 259

4.15.1   Intermezzo: Topological minors 259

4.15.2   Strong immersions in series - parallel digraphs 261

4.15.3   Intermezzo on 2 - trees 262

4.15.4   Series parallel - triples 263

4.15.5 A well quasi - order for one way series parallel
- triples 268

4.15.6 Series parallel separations 270

4.15.7 Coda 275

4.15.8 Exercise 281

4.16 Asteroidal sets 282

4.16.1 AT - free graphs 283

4.16.2 Independent set in AT-free graphs 283

4.16.3 Exercise 285

4.16.4 Bandwidth of AT-free graphs 286

4.16.5 Dominating pairs 290

4.16.6 Antimatroids 290

4.16.7 Totally balanced matrices 292

4.16.8 Triangle graphs 295

4.17 Sensitivity 296

4.17.1 What happened earlier 297

4.17.2 Cauchy's interlace lemma 298

4.17.3 Hypercubes 298

4.17.4 Möbius inversion 300

4.17.5 The equivalence theorem 301

4.17.6 Further reading 304

4.18 Homomorphisms 304

4.18.1 Retracts 305

4.18.2 Retracts in threshold graphs 306

4.18.3 Retracts in cographs 307

4.19 Products 312

4.19.1 Categorical products of cographs 313

4.19.2 Tensor capacity 314

4.19.3 Cartesian products 317

4.19.4 Independence domination in cographs 318

4.19.5 $\theta_e(K_n \times K_n)$ 319

4.20 Outerplanar Graphs 321

4.20.1 k – Outerplanar Graphs 322

4.20.2 Courcelle's Theorem 323

4.20.3 Approximations for Planar Graphs 323

4.20.4 Independent Set in Planar Graphs 323

4.21 Graph isomorphism 325

Bibliography   **327**

Index   **335**

Preface

A series of summerschools organized by the University of Science and Technology in China germinated this book. It is a collection of texts that are either 'trend-setters' or just good examples. A first draft of this text was prepared in 2016 and 2017. It was substantially updated in subsequent years.

This book provides an introduction to the research area of graph algorithms and reviews the development over the last decade. The contents is divided into parts titled 'Graphs,' 'Algorithms,' 'Problem Formulations,' and 'Recent Trends.' The first part reviews some graph - theoretic concepts. The second part presents a few early highlights in graph algorithms. The third part is a very short introduction into graph algebras and monadic second order - logic. The last part of the book uses 'treewidth' as a stepping stone and talks through a wide variety of recent trends.

The book assumes familiarity with a few basic concepts and programming techniques that are taught during a first year computer science course in algorithms. It provides a smooth introduction for those who want to dive deep into this fascinating research area. The book contains a lot of exercises, many up at present - day research - level; they will keep the reader pleasantly entertained for many hours.

About the authors

Dr. Mingyu Xiao is a professor in the school of computer science and engineering, University of Electronic Science and Technology of China, Chengdu, China. He received his PhD in Computer Science from the Chinese University of Hong Kong in 2008.

Dr. Ton Kloks is a researcher of graph algorithms. He studied mathematics at Eindhoven University in the Netherlands during the 1980s. He received his PhD in Computer Science from Utrecht University in The Netherlands in 1993 for his thesis on treewidth.

Acknowledgments

AN INVITATION by Dr. Mingyu Xiao, to introduce second and third year students of his department during a one - week - course to graph algorithmics, led me — Ton Kloks — to write a first draft of this text that lies before you now. I am grateful to my wife, to Mingyu Xiao, and to the University of Electronic Science and Technology of China for giving me this opportunity.

THIS TEXT originated in 2016 – 2017, and was substantially updated during the summers of 2018, 2019, 2020 and 2021.

Graphs

THIS BOOK provides an introduction to the research area of graph algorithms. — To start — in this chapter we review some graph–theoretic concepts.

Figure 1.1: Kloks' Teacher, Professor Jaap Seidel, once told him that this was his favorite graph. (The point in the middle was added on only when he grew older.)

Definition 1.1. A graph G is an ordered pair of finite sets

$$G = (V, E).$$

The elements of the nonempty set V are called points — or vertices.[1] The set E is a subset of the unordered pairs of points

$$E \subseteq \{\{a, b\} \mid \{a, b\} \subseteq V\}$$

and the elements of E are called lines — or edges.[2]

[1] We have: one vertex, and two vertices.

[2] An edge is a set with two elements, which are two vertices.

When the sets V and E of a graph G are not clear from the context we denote them as $V(G)$ and $E(G)$. We require that $V \neq \varnothing$.[3] A graph is called empty if $E = \varnothing$.

[3] A structure with $V = \varnothing$ is referred to as a 'null - graph.' (It is not a graph.) In English: 'null' = 'invalid.'

The two vertices of an edge are called the endpoints of that edge. An endpoint of an edge is said to be 'incident' with that edge. Two vertices x and y are adjacent if $\{x, y\} \in E$ and they are nonadjacent if $\{x, y\} \notin E$. When two vertices are adjacent — we say that they are — neighbors — of each other.

All graphs that we consider in this book are finite — that is — the set V is a nonempty finite set.

Definition 1.2. A set V is <u>finite</u> if

$$\exists_{k \in \mathbb{N}} \ |V| \leqslant k.$$

By the way, we use

$$\mathbb{N} = \{ 1, \ 2, \ \dots \}.$$

The set $\mathbb{N}$ of the <u>natural numbers</u> is, by definition, countable but, it is <u>not</u> finite.

1.1   Isomorphic Graphs

Definition 1.3. Two graphs G_1 and G_2 are <u>isomorphic</u> if there exists a bijection $\pi : V(G_1) \to V(G_2)$ satisfying

$$\{x, y\} \in E(G_1) \quad \Leftrightarrow \quad \{\pi(x), \pi(y)\} \in E(G_2). \qquad (1.1)$$

Exercise 1.1

Show that 'being isomorphic' is an equivalence relation — that is to say — show that

1. Every graph is isomorphic to itself

2. If G_1 is isomorphic to G_2 then G_2 is isomorphic to G_1

3. If G_1 is isomorphic to G_2 and G_2 is isomorphic to G_3 then G_1 is isomorphic to G_3.

1.2   Representing graphs

A graph can be represented by its <u>adjacency matrix</u>. Let G be a graph. The adjacency matrix A of G is a symmetric $0/1$−matrix with rows and columns corresponding to the vertices of G. The elements $A_{x,y}$ of A are defined as follows.

$$\forall_{x \in V} \ \forall_{y \in V} \quad A_{x,y} = \begin{cases} 1 & \text{if } \{x, y\} \in E(G), \text{ and} \\ 0 & \text{otherwise.} \end{cases} \qquad (1.2)$$

— For example — the graph in Figure 1.1 on Page 1 has an adjacency matrix:

$$\begin{pmatrix} 0 & 0 & 0 & 0 & 0 & 0 \\ 0 & 0 & 1 & 0 & 0 & 1 \\ 0 & 1 & 0 & 1 & 0 & 0 \\ 0 & 0 & 1 & 0 & 1 & 0 \\ 0 & 0 & 0 & 1 & 0 & 1 \\ 0 & 1 & 0 & 0 & 1 & 0 \end{pmatrix} \tag{1.3}$$

1.3 Neighborhoods

Definition 1.4. The neighborhood $N(x)$ of a vertex $x \in V(G)$ is the set of vertices that are adjacent to x — that is

$$N(x) = \{ y \in V \mid \{ x, y \} \in E \}. \tag{1.4}$$

When the graph G is not clear from the context we use $N_G(x)$ instead of $N(x)$.

The degree of a vertex x is the number of its neighbors, ie [4]

$$d(x) = |N(x)|. \tag{1.5}$$

We use the notation

$$N[x] = N(x) \cup \{x\} \tag{1.6}$$

to denote the 'closed neighborhood' of a vertex x. For a nonempty set W of vertices we write

$$N(W) = \left(\bigcup_{w \in W} N(w) \right) \setminus W$$

$$= \{ y \mid y \in V \setminus W \ \text{and} \ N(y) \cap W \neq \varnothing \}. \tag{1.7}$$

We use $N[W] = N(W) \cup W$ to denote the closed neighborhood of W.

[4] A graph G is called regular if all vertices have the same degree.

Figure 1.2: The Petersen graph is regular with degree 3.

1.4   Connectedness

A graph is connected if one can walk from any point to any other point via the edges of the graph.[5]

Definition 1.5. A graph G is <u>connected</u> if $|V| = 1$ — or else — if for every <u>partition</u> — $\{A, B\}$ of V — there exists elements $a \in A$ and $b \in B$ satisfying $\{a, b\} \in E$.

A graph is <u>disconnected</u> if it is not connected.

A partition of a set is defined as follows.

Definition 1.6. A collection of sets $\{V_1, \ldots, V_t\}$ is a <u>partition</u> of a set V if

1. $t \geqslant 2$,

2. all $V_i \neq \varnothing$,

3. for all $i \neq j$, $V_i \cap V_j = \varnothing$, and

4. $\cup V_i = V$.

[5] When I told my teacher AB, back in 1984, that I needed this property to finish my proof, he said: "That's called CONNECTED!"

1.5   Induced Subgraphs

Let G be a graph. A <u>subgraph</u> of G is a graph H with [6]

$$V(H) \subseteq V(G) \quad \text{and} \quad E(H) \subseteq E(G). \tag{1.8}$$

Definition 1.7. Let G be a graph. Let $W \subseteq V(G)$ and let $W \neq \varnothing$. The subgraph of G <u>induced</u> by W is the graph H with

$V(H) = W$ and

$E(H) = \{\{a, b\} \mid \{a, b\} \subseteq W \text{ and } \{a, b\} \in E(G)\}.$

[6] The graph H is a <u>spanning subgraph</u> of G if $V(H) = V(G)$ and $E(H) \subseteq E(G)$. When H is a spanning subgraph of G and also a tree it is called a spanning tree of G. Clearly, a graph can only have a spanning tree if it is connected.

For a graph G and a nonempty set $W \subseteq V(G)$ we denote the subgraph of G induced by W as $G[W]$. We also write — where we use $V = V(G)$ —

$$G - W = G[V \setminus W], \quad \text{when } V \setminus W \neq \varnothing. \qquad (1.9)$$

For a vertex x we write $G - x$ instead of $G - \{x\}$. For an edge $e = \{x, y\}$ we write $G - e$ for the graph with vertices $V(G)$ and edges $E(G) \setminus \{e\}$.

1.6   Paths and Cycles

Let G be a graph. A path[7] in G is a nonempty set of vertices that is ordered such that consecutive pairs are adjacent. A path has — by definition — at least one vertex. To denote a path P in G we use the notation

[7] In a path all vertices are distinct.

$$P = [x_1 \quad \cdots \quad x_t]. \qquad (1.10)$$

— Here — the vertices of P are

$$V(P) = \{x_1, \cdots, x_t\} \subseteq V(G).$$

The edges of P are the pairs of vertices that are consecutive in the ordering. Edges in G — that connect vertices in P which are not consecutive — are called chords of P.

A path with n vertices has $n - 1$ edges.

The points x_1 and x_t are the endpoints of P and we also say that P runs between x_1 and x_t. We call a path P an $x \sim y$–path if x and y are the endpoints of P. The length of P is the number of edges in it — that is to say —

$$\ell(P) = |E(P)| = |V(P)| - 1.$$

Definition 1.8. For any two vertices x and y in a graph their distance is defined as

$$d(x, y) = \min\{|E(P)| \mid P \text{ is an } x \sim y \text{ –path}\}. \qquad (1.11)$$

In the case where $V(P) = V(G)$ and $E(P) = E(G)$ — the graph G is called a path.[8] We use a special notation to denote graphs that are paths with n vertices — namely —

$$\forall_{n \in \mathbb{N}} \quad P_n \text{ is the path with } n \text{ vertices.}$$

Exercise 1.2

Design an algorithm to check whether a graph G given as input is a path.

A cycle [9] C in a graph G is a set of at least three vertices that is ordered

$$C = [c_1 \quad \cdots \quad c_t] \tag{1.12}$$

such that any two consecutive vertices are adjacent in G and — furthermore — also the first and last vertex of the sequence are adjacent.

The edges of the cycle C are the consecutive pairs and the pair $\{c_1, c_t\}$. Edges

$$\{c_i, c_j\} \in E(G) \setminus E(C)$$

are called <u>chords</u> of C.

When $V(G) = V(C)$ and $E(G) = E(C)$ we say that G is a cycle. For graphs isomorphic to a cycle with n vertices we use the special notation

$$C_n \text{ is the cycle with } n \text{ vertices.}$$

Exercise 1.3

Show that a graph G is regular with all vertices of degree two, if and only if every component of G induces a cycle.

Exercise 1.4

Show that $\bar{C}_5$ and C_5 are isomorphic. Are there other cycles isomorphic to their complement?

[8] A graph is a path if it contains a path to which it is isomorphic.

Figure 1.3: The figure shows a path with 4 vertices; that is, P_4.

[9] In a cycle all vertices are distinct.

Figure 1.4: C_5: a cycle with 5 vertices

1.7 Complements

Strawberries and cream
complement each other
nicely!

Let G be a graph. The <u>complement</u> of G is the graph $\bar{G}$ with the following sets of vertices and edges.

$V(\bar{G}) = V(G)$ and

$E(\bar{G}) = \{\{a, b\} \mid \{a, b\} \subseteq V(G) \text{ and } \{a, b\} \notin E(G)\}.$

Figure 1.5: The figure shows $\bar{C}_4$, the complement of the 4-cycle. Notice that this graph is disconnected.

1.8 Components

UNIONS of connected graphs — without any additional edges between them — create new graphs. The constituent parts of the new graph are called its components.

Definition 1.9. A <u>component</u> of a graph G is a maximal set of vertices $W \subseteq V(G)$ such that $G[W]$ is connected.

Figure 1.6: This is the house. It is the complement of P_5: $\bar{P}_5$.

Notice that, the set of components of a graph G forms a partition of $V(G)$ if and only if G is disconnected. Each component induces a connected subgraph of G.

1.8.1 Rem's Algorithm

Let G be a graph and let $V(G) = [n]$.[10] Rem's algorithm, Algorithm 1 computes a function $\delta : V(G) \to [n]$ satisfying

[10] $[n] = \{1, \ldots, n\}$

$$\forall_k \quad \delta(k) \leqslant k \quad \text{and}$$

(i) $\delta(k) = k$ if vertex k belongs to the component with number k

(ii) $\delta(k) < k$ if vertex k lies in the same component as vertex $\delta(k)$.

For the number of components we have

$$\# \text{ components of } G = \#\{\, x \mid x \in V \quad \text{and} \quad \delta(x) = x \,\}. \quad (1.13)$$

Notice that when G is the empty graph [11] the identity function is a solution — it puts a vertex k in the component with number k, for $k \in [n]$.

[11] G is empty if $E(G) = \varnothing$.

Rem's algorithm initializes the function δ as the identity function and, as it — one by one — adds the edges to the graph, it updates δ. The objective is to decrease the number of applications needed for a vertex to find the representative vertex in its component. [12] The algorithm is optimal in the sense that δ decreases at every pass of the loop that starts in Line 11.

[12] That is, the number of times one has to apply δ before it gets constant.

Exercise 1.5

Prove the correctness of Rem's algorithm in Algorithm 1. Show that it runs in $O(n^2)$ time. Implement and test it on the graph shown in Figure 2.2 on Page 23 — that is the graph with — say 20 — disjoint triangles.

Exercise 1.6

Design an algorithm that — using Rem's algorithm as a subroutine — checks whether two vertices x and y are in the same component of a graph G.
Hint: Analyze what happens to their components if you add the edge $\{x, y\}$ to the graph.

Exercise 1.7

Suppose we measure the 'efficiency' of the algorithm by the maximal number of times one has to apply the function δ to a vertex to find its component's number. The efficiency of Rem's algorithm seems to depend on the order in which we add the edges. Is there a clever choice?
Hint: Suppose G is a path. Analyze different orderings in which to add the edges.

```
 1: procedure REM
 2:
 3:     for i ∈ [n] do
 4:         δ(i) ← i
 5:     end for
 6:
 7:     for {p, q} ∈ E(G) do
 8:         p₀, q₀ ← p, q
 9:         p₁, q₁ ← δ(p₀), δ(q₀)
10:
11:         while p₁ ≠ q₁ do
12:             if p₁ < q₁ then
13:                 δ(q₀) ← p₁
14:                 q₀, q₁ ← q₁, δ(q₁)
15:             else
16:                 δ(p₀) ← q₁
17:                 p₀, p₁ ← p₁, δ(p₁)
18:             end if
19:         end while
20:
21:     end for
22: end procedure
```

1.9   Separators

Definition 1.10. Let G be a graph.

1. A <u>separator</u> is a set $S \subset V$ such that $G - S$ is disconnected.

2. For two nonadjacent vertices a and b, a set S is a minimal $a \mid b$-separator if a and b are in different components of $G - S$ and no proper subset of S has that property.

3. A set $S \subset V(G)$ is a <u>minimal separator</u> if S is a minimal $a \mid b$-separator for some nonadjacent pair a and b.

— In a graph — the empty set is a minimal separator if and only if the graph is disconnected. If a minimal separator contains only one vertex that vertex is called a cutvertex.

Exercise 1.8

Let a and b be nonadjacent vertices in a graph. Prove that there is exactly one minimal $a \mid b$-separator contained in $N(a)$.

Exercise 1.9

Show that a set S is a minimal separator in a graph G if and only if $G - S$ has two components that have a neighbor of every vertex in S.

Figure 1.7: Notice that one minimal separator can be properly contained in another one. This is clear when G is disconnected. In this example, the cutvertex is properly contained in a minimal separator for the nonadjacent pair of its neighbors in the 5-cycle.

1.10   Trees

Definition 1.11. A connected graph T is a <u>tree</u> if it contains no cycles — that is — if no induced subgraph of T is a cycle.

Remark 1.12. Notice that Definition 1.11 implies that every connected induced subgraph of a tree is a tree.

Figure 1.8: A tree

Definition 1.13. A graph is a forest if each of its components induces a tree.

Exercise 1.10

Prove or disprove: A connected graph is a tree if and only if every minimal separator in it has cardinality 1. What are graphs in which every connected induced subgraph with at least three vertices has a cutvertex?

Theorem 1.14. *A connected graph is a tree if and only if every connected induced subgraph with at least two vertices has a vertex of degree one.* [13]

Proof. When a connected graph is not a tree then it contains a cycle, and so, the graph has an induced subgraph — the cycle — without vertex of degree 1.

Let T be a tree. When $|V(T)| = 1$ the only induced subgraph of T is T itself. Since T has no induced subgraph with at least two vertices, the condition is void (as it should be).

Now assume that $|V(T)| > 1$ and let T be connected. We show that T has two leaves. Let x be any vertex of T and let $C_1, \cdots, C_t$ be the components of $T - x$. If some component C_i has only one vertex, then this vertex has degree one, since its only neighbor is x.

If x has two neighbors in C_1, say y and z, then there is a path P in $T[C_1]$ connecting y and z. Then $V(P) \cup \{x\}$ induces a cycle in T.

By induction — on the number of vertices of T — we may assume that each C_i induces a tree with at least two vertices. By induction on the number of vertices — since $|C_i| > 1$ — $T[C_i]$ contains at least <u>two</u> leaves. At least one of those leaves is not adjacent to x — and so — there is a leaf in $T[C_i]$ which is a leaf in T.

To see that T has at least two leaves — observe that when $T - x$ has only one component that component contains a leaf

[13] In a tree T we call the vertices of degree at most one the <u>leaves</u> of T. Note that we have: 'one leaf' and 'two leaves.' (I did it wrong in my PhD thesis ;-(.) In graphs, vertices of degree one are called <u>pendant</u> vertices. A pendant vertex is pendent (="dangling.") 'Pedant' means something entirely different.

and x itself is a leaf. When $T - x$ has at least two components
— each component contains a leaf — and so there are at least
two leaves.

This proves the theorem.   □

Notice that the proof of Theorem 1.14 shows that we can
prune [14] leaves off a tree, until no vertex remains.

Corollary 1.15. *A graph* T *is a tree if and only if it has
an* _elimination order_ *— that is an ordering of* $V(T)$ *— say*

$$[x_1 \quad \cdots \quad x_n]$$

such that

$$\forall_{1 \leqslant i \leqslant n} \quad G[V_i] \text{ is a tree and } x_i \text{ is a leaf in } G[V_i]$$
$$\text{where } V_i = \{x_i, \cdots, x_n\}. \quad (1.14)$$

Exercise 1.11

Design an algorithm that checks whether a graph given as
input is a tree.

1.11 Bipartite Graphs

Definition 1.16. A graph G is <u>bipartite</u> if $|V(G)| = 1$ or
else there is a partition of $V(G)$ — say $\{A, B\}$ — such that

$$\forall_{e \in E(G)} \quad e \cap A \neq \varnothing \text{ and } e \cap B \neq \varnothing. \quad (1.15)$$

The two sets A and B in a partition as above are called
<u>color classes</u> of G. Notice that these color classes are indepen-
dent sets in G — that is — no two vertices in one color class
are adjacent.

Definition 1.17. Let G be a graph and let

$$S \subseteq V \text{ and } S \neq \varnothing.$$

The set S is an <u>independent set</u> if no two vertices in S are
adjacent.

[14] 'To prune' means, to cut
off branches from a plant or
tree. In our case, we just pick
leaves one by one.

Exercise 1.12

Show that every tree is bipartite.

Exercise 1.13

A bipartite graph is
<u>complete bipartite</u> if every
pair of vertices in different
color classes is adjacent. We
denote a complete bipartite
graph, with a and b vertices
in its color classes, as $K_{a,b}$.
What are the minimal
separators in a complete
bipartite graph?

We represent the maximal cardinality of an independent set in G by

$$\alpha(G).$$

Let C be some set — the elements of which are called colors. A coloring of a graph G is a map

$$V(G) \to C$$

with the property that the endpoints of any edge in G receive different colors of C.

Definition 1.18. The <u>chromatic number</u> of a graph G

$$\chi(G)$$

is the smallest number of colors needed to color the vertices such that adjacent vertices have different colors.

By definition a graph G is bipartite if and only if $\chi(G) \leqslant 2$.

Exercise 1.14

Design an algorithm that checks whether a graph is bipartite. Hint: It doesn't seem clever to use Theorem 1.19.

We have seen that trees are exactly those connected graphs that contain no cycles. Bipartite graphs are characterized via cycles as follows.

Theorem 1.19. *A graph is bipartite if and only if all cycles in it are even.*

Proof. Let G be bipartite and let $\{A, B\}$ be a partition of V such that all edges have one end in A and the other in B. Let $C = [c_1 \cdots c_t]$ be a cycle. Then the elements c_i alternate between A and B. This proves that

$$|V(C)| = |E(C)| = t \text{ is even.} \tag{1.16}$$

Let G be a graph and assume that all cycles in G are even. We may assume that G is connected. Start coloring G by assigning an arbitrary vertex r the color A. Define the set A as the set of vertices that are at even distance from r. Let

$$B = V \setminus A.$$

We claim that $\{A, B\}$ is a coloring of G. Assume that A contains two adjacent vertices x and y. Then a shortest path from r to x — together with a shortest path from r to y — plus the edge $\{x, y\}$ — contains an odd cycle.

This proves the theorem.   □

1.12   Linegraphs

Definition 1.20. Let G be a graph with at least one edge. [15] The linegraph $L(G)$ of G is the graph

$$V(L) = E(G) \quad \text{and}$$
$$E(L) = \{\{e_1, e_2\} \mid \{e_1, e_2\} \subseteq E(G) \quad \text{and} \quad e_1 \cap e_2 \neq \varnothing\}. \tag{1.17}$$

[15] When a graph is empty, its 'linegraph' would have no vertices, which makes it not a graph.

Exercise 1.15

Describe the linegraphs of trees (with at least one edge).

Notice that linegraphs are graphs without claws (see Figure 2.11). For any fixed graph H, a graph is H-free if it does not have an induced subgraph isomorphic to H.

1.13   Cliques and Independent Sets

Definition 1.21. A clique in a graph G is a nonempty set

$$C \subseteq V$$

such that any two vertices of C are adjacent.

— So — a clique in G is an independent set in $\bar{G}$ and vice versa. [16] Let $\omega(G)$ denote the maximal cardinality of a clique in G. Then

$$\omega(G) = \alpha(\bar{G}).$$

We call a clique with three vertices a triangle. Of course, bipartite graphs have no triangles, since those are odd cycles. It follows that $\omega(G) \leqslant 2$ whenever G is bipartite.

A graph is called a clique if every pair of its vertices are adjacent. [17] When G is a clique we have

$$\omega(G) = |V(G)| = \chi(G) \quad \text{and} \quad \alpha(G) = 1.$$

— Similarly — a graph is an independent set if it has no edges. When G is an independent set we have

$$\omega(G) = \chi(G) = 1 \quad \text{and} \quad \alpha(G) = |V(G)|.$$

1.14   On Notations

— TO CONCLUDE THIS CHAPTER — when there is no confusion possible we use — when a generic graph G underlies the discussion

$$n = |V(G)| \quad \text{and} \quad m = |E(G)|.$$

— Similarly — we use

$$V = V(G) \quad E = E(G) \quad \omega = \omega(G) \quad \alpha = \alpha(G) \quad \&\text{tc.}$$

whenever it is clear and convenient. [18]

We freely abuse notation when it clarifies the text — for example — when S is some subset of vertices of a graph G then we use S also to denote $G[S]$ (see the Sidenote 2.4 on Page 30). The general rule applied removes uninformative variables and parentheses. [19]

[16] An independent set is defined in Definition 1.17.

[17] For a clique with n vertices we use the notation K_n.

[18] Perhaps I should have used v and e, instead of n and m, but this choice has the advantage that I am used to it!
I *should* have used G instead of V (and use the 'old' notation where G is synonymous with (G, E)), but, unfortunately, 'recent traditions' kept me back.

[19] ...to improve readability; we don't want to compress the text.   ;-)

Algorithms

— To get a taste of graph algorithms — it seems a good idea to have a look at the early achievements that are still widely in use today.

Before we start a word of warning.[1] Debugging may help you find errors in your algorithm, but it can never prove that your algorithm is correct.

— Actually [2] — there cannot be a procedure that tests if an algorithm terminates or not. Minsky gives the following proof of this. Assume there were a procedure proper($\cdot$) that takes as input any procedure L and outputs TRUE if L terminates and FALSE otherwise. The following example provides a contradiction.

Consider the procedure L shown in Algorithm 2. When L is proper then it is improper and vice versa. Hence, the procedure proper($\cdot$) can not exist.

```
1: procedure L
2:    use package proper
3:
4:        while proper(L) do
5:            whistle once
6:        end while
7: end procedure
```

Algorithm 2: Minsky's example of a procedure that is neither proper nor improper.

2.1   Finding and counting small induced subgraphs

TRIANGLES IN GRAPHS can be found via fast matrix multiplication of its adjacency matrix. This gives an algorithm to find a triangle in $O(n^\alpha)$ where $\alpha < 2.376$.

In 1997 Alon, Yuster and Zwick showed that a triangle can be found in $O(m^{2\alpha/\alpha+1}) = O(m^{1.41})$. (This improved an earlier $O(m^{3/2})$ algorithm by Itai and Rodeh.) [3]

In this section we show how their method can be used to find a diamond in a graph in $O(m^{3/2} + n^\alpha)$. [4]

[3] In this section we express the run - time of algorithms as a function of n and m.

[4] A diamond is a graph with 4 vertices; obtained from K_4 by removing one edge. A graph is diamond-free if no induced subgraph is isomorphic to the diamond.

Exercise 2.1

A graph is diamond - free if and only if the neighborhood of every vertex induces a graph in which every component is a clique.

Hint: If a graph has no diamond then every neighborhood is P_3-free — ie — every component of a neighborhood is a clique.

The algorithm to check if a graph G has a diamond first partitions the vertices in those that have 'low' degree and 'high' degree. Let D be some number. A vertex is low degree if its degree is at most D and otherwise it is high degree. Let L be the set of vertices that have low degree and let H be the set of vertices that have high degree.

The search for a diamond is split in 4 parts.

Phase 1.   check if G has a diamond with a vertex of degree 3 that is of low degree

Phase 2.   check if G has a diamond with a vertex of degree 2 that is of low degree

Phase 3.   if no diamond was found then remove all vertices of low degree. Let G^* be the graph that remains.

Phase 4.   check if G^* has a diamond.

We show how each phase is implemented. We assume that the adjacency matrix A is given. For each $x \in L$ construct adjacency lists for the graph induced by $N(x)$. This can be accomplished in $O(d(x)^2)$ time. Then compute the components of $G[N(x)]$ and check if each component is a clique. This can be done in $O(d(x)^2)$ time. If some component is not a clique then a P_3 is found in $O(d(x)^2)$ time. [5] It follows that this phase can be completed in time

[5] Exercise !

$$\sum_{x \in L} d(x)^2 \quad \leqslant \quad 2 \cdot D \cdot m.$$

We describe the implementation of Phase 2. Let $x \in L$ and let C be a maximal clique in $N(x)$. For each pair $y, z \in C$ check if

$$A^2_{y,z} \quad > \quad |C| - 1.$$

If that is the case then y and z have a common neighbor outside $N[x]$ — ie — we find a diamond. The diamond can be produced in linear time when all adjacency lists are sorted. We leave it as an exercise to check that Phase 2 runs in $O(D \cdot m + n^\alpha)$.

Assume that Phase 1 and Phase 2 do not produce a diamond. Notice that

$$V(G^*) \quad \leqslant \quad \frac{2 \cdot m}{D}.$$

Repeat the procedure described in Phase 1 for all vertices of G^*. This can be implemented to run in

$$\sum_{x \in H} d_H(x)^2 \quad = \quad O\left(m \cdot |H|\right) \quad = \quad O\left(\frac{m^2}{D}\right).$$

Theorem 2.1. *There exists an algorithm that finds a diamond in a graph if there is one. With the adjacency matrix of the graph as input the algorithm runs in $O(n^\alpha + m^{3/2})$.*

Proof. BY THE ABOVE the total run - time is at most

$$O\left(D \cdot m + \frac{m^2}{D} + n^\alpha\right).$$

Choose $D = \sqrt{m}$. □

Exercise 2.2

Use a similar technique to show there is an algorithm to check if a
connected graph is claw-free that runs in $O(m^{(\alpha+1)/2}) = O(m^{1.69})$.
6

[6] The claw is shown in Figure 2.11 on Page 84. It is a tree with 4 vertices of which 3 are leaves.

HINT: If a graph is claw-free then every vertex has at most $2\sqrt{m}$
neighbors. When every vertex has at most $2\sqrt{m}$ neighbors then
do a fast matrix multiplication for each neighborhood and check
for a $\bar{K}_3$. This step can be performed in time proportional to

$$\sum_x d(x)^\alpha \;\leqslant\; (2\sqrt{m})^{\alpha-1} \cdot \sum_x d(x) \;\leqslant\; 2^\alpha \cdot m^{(\alpha+1)2}.$$

2.2   Bottleneck domination

Let G be a graph and let $w : V \to \mathbb{R}$ be a function which assigns
to every vertex x a weight $w(x)$. We assume that arithmetic
operations on vertex weights can be performed in $O(1)$ time.

Definition 2.2. Let (G, w) be a weighted graph. For $W \subseteq V$ the
<u>bottleneck</u> of W is

$$\max\{w(x) \mid x \in V\}.$$

Definition 2.3. Let G be a graph. A set $D \subseteq V$ is a <u>dominating
set</u> if every vertex of $V \setminus D$ has a neighbor in D.

For example, every maximal
independent set in a graph
is a dominating set.

The <u>bottleneck domination problem</u> is the following.
Input: A weighted graph (G, w).
Output: A dominating set with minimal bottleneck.

T. Kloks, D. Kratsch, C. Lee and J. Liu, *Improved bottleneck domination algorithms*, Discrete Applied Mathematics **154** (2006), pp. 1578–1592.

Exercise 2.3

Prove the following theorem.

Theorem 2.4. *There exists a linear time - algorithm to solve the bottleneck domination problem.*

HINT: Let (G, w) be a weighted graph. For $x \in V$ let

$$m(x) \quad = \quad \min\{w(y) \mid y \in N[x]\}.$$

Let

$$\rho \quad = \quad \max\{m(x) \mid x \in V\}.$$

Show that the minimal bottleneck is ρ.

Definition 2.5. Let G be a graph. A <u>total dominating set</u> is a set $D \subseteq V$ such that every vertex of V has a neighbor in D. [7]

[7] If the graph has isolated vertices then there is no total dominating set.

Exercise 2.4

Show that D is a total dominating set if it is a dominating set and $G[D]$ has no isolated vertices.

Exercise 2.5

Prove the following theorem.

Theorem 2.6. *There exists a linear time - algorithm to compute a total dominating set with smallest bottleneck in a weighted graph.*

HINT: For $x \in V$ define

$$m'(x) \quad = \quad \min\{w(y) \mid y \in N(x)\}$$

and let

$$\rho' \quad = \quad \max\{m'(x) \mid x \in V\}.$$

Show that the smallest bottleneck of a total dominating set is ρ'.

2.3 The Bron & Kerbosch Algorithm

In this section we'll have a look at the Bron–Kerbosch algorithm. It was developed in the early 1970s at Eindhoven's University of Technology in The Netherlands. It remains to this day a winner.

Recall the definition of a clique (Definition 1.21).[8]

Definition 2.7. A clique C is <u>maximal</u> if

$$\forall_{x \notin C} \exists_{y \in C} \ \{x, y\} \notin E. \qquad (2.1)$$

For a graph G let

$$\Omega(G)$$

represent the set of maximal cliques in G.

[8] A clique in a graph G is a set of vertices that are all pairwise adjacent.

Figure 2.1: A clique is maximal if it is not contained in a larger one. Obviously, this doesn't mean that there <u>is</u> no larger one! This graph is called the net. Edges incident with pendant vertices are maximal cliques. The largest maximal clique is the triangle. A largest maximal clique is called a <u>maximum</u> clique.

The algorithm of Bron and Kerbosch lists all the maximal cliques in a graph. In this section we describe a variant of the original Bron and Kerbosch –algorithm and we analyze its time complexity.

When a graph G has only vertices of degree at most 2 then each component of G is a path or a cycle. In that case the number of maximal cliques is at most n. It seems that when a graph has a lot of maximal cliques it has a lot of vertices of high degree.

Let's look at the case where all vertices of G have degree at least $n - 3$. We say that G is <u>high degree</u>.

Exercise 2.6

Show that a graph G is high degree if and only if every induced subgraph of G is that.

Figure 2.2: This figure show the complement of a high degree-graph. For this graph we use notations

$$K_3 + \cdots + K_3 = t \cdot K_3,$$

to be read as 'a union of triangles.'

Notice that, when G is high degree, every component of $\bar{G}$ is a path or a cycle. Consider the case where $\bar{G}$ is a path — say

$$[x_1 \quad \cdots \quad x_n].$$

We can list the maximal cliques of G as follows. When $n \leqslant 3$ this is easy — so we assume henceforth that $n \geqslant 4$.

For the set of maximal cliques in $\bar{P}_n$ we have

$$\Omega(1 \cdots 1) = \{\{x_1\}\}, \quad \Omega(1 \cdots 2) = \{\{x_1\}, \{x_2\}\}, \quad \text{and}$$
$$\Omega(1 \cdots 3) = \{\{x_1, x_3\}, \{x_2\}\}.$$

and for $n \geqslant 4$,

$$\Omega(1 \cdots n) = \{\{x_n\} \cup C \mid C \in \Omega(1 \cdots n-2)\}$$
$$\bigcup \ \{\{x_{n-1}\} \cup C \mid C \in \Omega(1 \cdots n-3)\}. \quad (2.2)$$

Exercise 2.7

Implement this algorithm to list all maximal cliques in the complement of a path.

For the number of maximal cliques in $\bar{P}_n$ we have the recurrence

$$P(1) = 1 \quad P(2) = 2 \quad P(3) = 2 \quad \text{and}$$
$$P(n) = P(n-2) + P(n-3) \quad \text{for } n \geqslant 4,$$

where $P(n) = |\Omega(\bar{P}_n)|$. This solves as $O\left(\left(\frac{4}{3}\right)^n\right)$. (The architect Dom van der Laan considered this the ideal fraction of measurements for his buildings.)

Exercise 2.8

Design a similar algorithm to list all the maximal cliques in the complement of a cycle.

Hint: Can you adapt the algorithm for paths?

Some straightforward calculations show that the high–degree graph — with the most maximal cliques — is a graph G — satisfying one of the following.

i. If $n = 0 \bmod 3$, all components of $\bar{G}$ are triangles, that is, G has $3^{n/3}$ maximal cliques.

ii. If $n = 1 \bmod 3$ and $n > 1$, all components of $\bar{G}$, except two, are triangles, and the two exceptional components are edges.[9] This gives $4 \cdot 3^{(n-4)/3}$ maximal cliques if $n > 1$. We have one maximal clique if $n = 1$.

iii. If $n = 2 \bmod 3$, all components of $\bar{G}$ — except one — are triangles, and the exceptional component is a single edge. This case gives $2 \cdot 3^{(n-2)/3}$ maximal cliques.

It follows that every high degree–graph has at most $3^{n/3}$ maximal cliques.

Let us first show that — indeed — among all graphs these graphs are the ones with the most maximal cliques.

Lemma 2.8. *Let* G *be a graph which is not high degree. Then*

$$|\Omega| \leqslant \mu \cdot 3^{n/3} \qquad (2.3)$$

for some $\mu \in \mathbb{R}$, $0 < \mu < 1$. *Here* $n = |V(G)|$.

Proof. The graph must have a vertex x of degree at most $n - 4$. Partition the set of maximal cliques into those that contain x and those that do not contain x. The first set of cliques are exactly the maximal cliques contained in $G[N(x)]$, with x added on to each of them.[10] The second set of cliques are maximal cliques in $G - x$. Possibly this second set contains

[9] The two exceptional components induce the complement of a 4-cycle with 4 maximal cliques.

Exercise 2.9

Part I

Consider the high degree-graph G in Figure 2.2. This graph is the complement of $\frac{n}{3}$ triangles. Show that

$$|\Omega(G)| = \left(3^{1/3}\right)^n$$
$$= (1.442\cdots)^n.$$

Part II

Show that the graph above has the most maximal cliques among all high degree-graphs.

[10] If $N(x) = \varnothing$, the vertex x is an <u>isolated vertex</u>. In that case the only maximal clique that contains x is $\{x\}$.

cliques that are not maximal in G — but — as an upperbound we get

$$|\Omega(G)| \leqslant \begin{cases} |\Omega(G[N(x)])| + |\Omega(G-x)| & \text{If } N(x) \neq \varnothing \\ 1 + |\Omega(G-x)| & \text{otherwise.} \end{cases} \quad (2.4)$$

By induction on the number of vertices in the graph

$|\Omega(G[N(x)])| \leqslant 3^{(n-4)/3}$ since x has degree at most $n-4$.

By induction also

$$|\Omega(G-x)| \leqslant 3^{(n-1)/3}.$$

It follows that [11]

$$\begin{aligned} |\Omega(G)| &\leqslant 3^{(n-4)/3} + 3^{(n-1)/3} \\ &= (3^{-4/3} + 3^{-1/3}) \cdot 3^{n/3} \\ &= 3^{-4/3} \cdot (1+3) \cdot 3^{n/3} \\ &= 3^{-4/3} \cdot 4 \cdot 3^{n/3} \\ &< 3^{n/3} \quad \text{since } 3^{4/3} > 4.326. \end{aligned}$$

This completes the proof.   □

[11] Since G is not high degree we may assume $n \geqslant 4$. When x is isolated, we have, for some $\mu \in (0,1)$, since $n \geqslant 4$,

$$|\Omega| \leqslant 1 + 3^{(n-1)/3}$$
$$\leqslant \mu \cdot 3^{n/3}.$$

Remark 2.9. I agree to it that the notation in (2.4) is *awful* with all those (useless) brackets! One of my teachers, Professor De Bruijn used to say: " Our notation for functions is terrible. — Unfortunately — it's not bad enough for people to feel the need to change it; so we're stuck with it and we'd better get used to it ! "

Lemma 2.10. *There is an algorithm that lists all the maximal cliques in a high degree graph in* $O(n^2 \cdot |\Omega|)$ *time.*

Proof. There is an algorithm that runs in $O(n^2)$ time and that computes the components of $\bar{G}$.[12] Since G is high degree,

[12] We assume that G is suitably represented.

each component of $\bar{G}$ is a path or a cycle, and we can find a suitable vertex ordering in each of these components.

The technique — explained on Page 23 ff— shows that there is an algorithm, that lists all cliques in the complement of a path or cycle in time $O(|W| \cdot N)$, where N is the number of maximal cliques in $G[W]$ for a component W of $\bar{G}$.

Let $\{W_1, \cdots, W_k\}$ be the set of components of $\bar{G}$. Then

$$\Omega(G) = \{ S_1 \cup \cdots \cup S_k \mid$$
$$\forall_i \quad S_i \text{ is a clique in the component } G[W_i] \}. \quad (2.5)$$

It is now straightforward to show that all maximal cliques in G can be listed in $O(n^2 \cdot |\Omega(G)|)$ time. $\square$

Exercise 2.10

Implement and run the algorithm described above for C_5.

Let A be the algorithm described above, that lists all maximal cliques in graphs that are high degree in

$$O(n^2 \cdot |\Omega|) = O\left(n^2 \cdot 3^{n/3}\right) \quad \text{time } (\text{as } n \to \infty).$$

We describe the algorithm of Bron and Kerbosch blow — in Algorithm 3. The algorithm to list all the maximal cliques consists of a call to the procedure with parameters:

$$B \& K(\varnothing, V, \varnothing).$$

An <u>invariant</u> is a property that holds true for the parameters of a procedure, at every call to it. An invariant is useful when the termination condition together with the invariant yields the desired solution, ie, the 'post condition.' The concept of an invariant of a procedure was introduced by Dijkstra in the early 1960s. As a concept it is a useful tool to prove the correctness of programs.

In the case of the procedure $B \& K$ described above the parameters can be described by the following invariant:

When we estimate time-bounds of graph algorithms, we are usually interested in the case where $n \to \infty$. We usually omit this addendum.

Choose=pick=select

Algorithm 3: The Bron–Kerbosch Algorithm

```
1: procedure B & K( R, P, X )
2:    if  P ∪ X = ∅ then  report  R
3:    else
4:        if  P = ∅ then  skip
5:        else
6:            if  G[P] is high degree then
7:                Compute  Ω( G[P] )  using  Algorithm A  and
8:                extend them with  R
9:            else
10:               Choose  x ∈ P  such that  |N(x) ∩ P| < |P| − 4
11:               B & K( R ∪ {x}, P ∩ N(x), X ∩ N(x) )
12:               B & K( R, P \ {x}, X ∪ {x} )
13:           end if
14:       end if
15:   end if
16: end procedure
```

1. $R = \emptyset$ or R is a clique in G.

2. P and X are disjoint sets and

$$P \cup X = \{ y \mid R \subseteq N(y) \}$$

— that is — the set $P \cup X$ contains those vertices $y \in V \setminus R$ such that $\{y\} \cup R$ is a clique.

Notice that — by virtue of the invariant — the set R is a <u>maximal</u> clique exactly when $P \cup X = \emptyset$; thence the Report command in Line 2.

The set P is called the set of candidates. When x is a candidate, chosen in Line 10, the algorithm lists all maximal cliques that contain $R \cup \{x\}$. Then x is removed from the set of candidates and put into the set X to maintain the invariant. — Finally — the remaining set of maximal cliques, that is, those that do not contain x, are listed via the call B & K with parameters

$$R, \quad P \setminus \{x\}, \quad \text{and} \quad X \cup \{x\}.$$

Exercise 2.11

Implement and run the Bron –Kerbosch algorithm for the 5-wheel W_5 in Figure 2.3 and for the Petersen graph (Figure 1.2 on Page 3).

Figure 2.3: The 5-wheel W_5

2.3.1   A Timebound for the B & K –Algorithm

We have seen that when G is high–degree, its maximal cliques can be listed using algorithm A in $O\left(n^2 \cdot 3^{n/3}\right)$ time. The sets P, X and R in the Bron and Kerbosch–algorithm may be implemented by a pointer structure or an array.

Theorem 2.11. *The Bron –Kerbosch algorithm runs in*

$$O(n^2 \cdot 3^{n/3}) \qquad time.$$

Proof. Let $t(n)$ denote the time needed to list all maximal cliques in a graph with n vertices. Write

$$t(n) = t^*(n) + r(n),$$

where $r(n)$ is the time spent on reporting maximal cliques. Then

$$r(n) = O\left(n \cdot |\Omega|\right).$$

The variant of the Bron–Kerbosch algorithm, in which each report statement is replaced by an $O(1)$ statement — like increasing a counter to count the number of maximal cliques — has then running time $t^*(n)$.

When $G[P]$ is high degree we have, by Lemma 2.10,

$$t^*(p) = O\left(3^{p/3}\right) \qquad where \ p = |P|. \tag{2.6}$$

We claim that — when $G[P]$ is not high degree

$$t^*(p) \leqslant t^*(p-4) + t^*(p-1) + O\left(p^2\right). \tag{2.7}$$

The term $t^*(p-4)$ corresponds to the case where a candidate x of degree at most $p-4$ is added to R. The search space reduces to $N(x)$, which has at most $p-4$ vertices.

The term $t^*(p-1)$ is the time needed to count the maximal cliques that do not contain x.[13] — In that case — the search space is $P-x$, ie, a graph with $p-1$ vertices.

The term $O(p^2)$ is the time needed to check if $G[P]$ is high degree, to update the sets R, P and X, and to find a candidate $x \in P$ with at most $p-4$ neighbors in P.

The recurrence (2.7) is similar to what we obtained for the number of maximal cliques in 2.4 on Page 25. It is readily checked that

$$t^*(n) = O\left(n^2 \cdot 3^{n/3}\right).$$

This completes the proof.   □

> [13] That is, there must exist a vertex $y \in P \setminus N_P[x]$ that is adjacent to all vertices of R.

> IMPORTANT: Make sure that you know how to solve a recurrence as in (2.7). If not, ask your teacher or checkout Concrete Math!

Remark 2.12. After having gathered a list of all the maximal cliques in the graph $G[N[x]]$, the remaining maximal cliques must contain at least one nonneighbor of x. — So — instead of finding all maximal cliques in $G-x$, we could list all maximal cliques that contain some nonneighbor of x. The analysis of Tomita et al. shows that this is 'much of muchness' — it does not improve the worst–case time estimate.

Exercise 2.12

This idea can be built into the algorithm by replacing the second recursive call (in Line 12) by a loop during which all nonneighbors of x are tried. In order to minimize the number of recursive calls Tomita et al. choose a vertex x with the <u>most</u> neighbors.

Eppstein et al. analyze the complexity for sparse graphs in terms of their 'degeneracy.'[14]

> [14] A graph is <u>k-degenerate</u> if every induced <u>subgraph</u> has a vertex of degree at most k.

Remark 2.13. For the maximal number of maximal cliques that a graph with $n > 1$ vertices may have, Moon and Moser derived the following formula.

$$g(n) = \begin{cases} 3^{n/3} & \text{if } n = 0 \bmod 3 \\ 4 \cdot 3^{\lfloor n/3 \rfloor - 1} & \text{if } n = 1 \bmod 3 \\ 2 \cdot 3^{\lfloor n/3 \rfloor} & \text{if } n = 2 \bmod 3. \end{cases} \quad (2.8)$$

For a slightly different proof see also Vatter.

The graph in Figure 2.2 is the unique [15] graph with n vertices and $3^{n/3}$ maximal cliques (when $n = 0 \bmod 3$).

[15] up to isomorphism

2.4   Total Order!

A binary relation on a set P is a subset of the set of ordered pairs in P, ie, a subset of the Cartesian product $P^2 = P \times P$ where

$$P \times P = \{ (q, r) \mid q \in P \text{ and } r \in P \}.$$

Definition 2.14. A <u>partial order</u> — or poset — P is a pair

$$(P, \leqslant),$$

where P is a set and $\leqslant$ is a binary relation on P satisfying:

When P is a poset, we use the same symbol P also for its set of elements. This abuse of notation was also common in graph theory. De Bruijn used to say about this: " As long as you know what you're talking about, there's no problem at all ! "

$$\begin{array}{lll} \forall_{x \in P} & x \leqslant x & \leqslant \text{ is reflexive,} \\ \forall_{x \in P} \forall_{y \in P} & (x \leqslant y \,\&\, y \leqslant x) \Rightarrow x = y & \text{antisymmetric,} \\ \forall_{x \in P} \forall_{y \in P} \forall_{z \in P} & (x \leqslant y \,\&\, y \leqslant z) \Rightarrow x \leqslant z & \text{and transitive.} \end{array}$$

Notice that possibly there are elements $x \in P$ and $y \in P$ for which neither $x \leqslant y$ nor $y \leqslant x$ holds.

Definition 2.15. A partial order $(P, \leqslant)$ is a <u>total order</u> if every pair of its elements are related by $\leqslant$. [16]

The TOTAL ORDERING PROBLEM is the problem to find a total order of a set V satisfying a collection of 'betweenness constraints.' We are given a collection R of ordered triples

$$(a, b, c) \in V^3.$$

The total order $\leqslant$ should satisfy:

$(a, b, c) \in R \quad \Rightarrow \quad (a < b < c) \quad \text{or} \quad (c < b < a),$

where we use $p < q$ to denote that $p \leqslant q$ and $p \neq q$.

$$(2.9)$$

Paraphrased — the required total order $\leqslant$ puts b 'between' a and c.

Let's look at the SIMPLE TOTAL ORDERING PROBLEM first. That problem is similar to the one above — except that each betweenness constraint has the form

$(a, b, c) \in R \quad \Rightarrow \quad a < b < c. \qquad (2.10)$

In the simple total ordering problem the collection of constraints builds a poset on the elements of V. In other words, the betweenness constraints define arcs between elements of V, say $x \to y$ if some betweenness relation implies that $x < y$.

Exercise 2.13

A simple total ordering problem on V has a solution if and only if the directed graph — defined above — is a DAG.

A <u>topological sort</u> of a digraph is a total ordering of its vertices such that for every arc $x \to y$ the vertex x comes before y in the total order.

In 1962 Kahn described an algorithm that finds a topological sort in a DAG — say $G = (V, A)$ — in time $O(n + m)$. [17]

[16] For example, $[n]$ and $\mathbb{N}$ are total orders. A total order is also called a <u>linear order</u>.

'constraint' = 'restriction'

A <u>digraph</u> is a graph in which each edge $\{x, y\}$ has a direction, either $x \to y$ or $y \to x$. A digraph is not a graph! If a digraph has no directed cycles, it is called a DAG, a directed, acyclic graph.

[17] $n = |V|$ and $m = |A|$

In Kahn's algorithm, Algorithm 4 on Page 32, the set S is the set of 'start–nodes,' which is the set of vertices without incoming arcs.[18] The set L contains the final linear ordering.

[18] Start-nodes are 'sources.' The vertices without outgoing arcs are called 'sinks.'

Algorithm 4: Kahn's Topological Sort

```
1: procedure KAHN( G = ( V , A ) )
2:
3:     L ← ∅
4:     S ← { x ∈ V | ∀_{y∈V} ¬ ( y → x ) }   ▷ S is the set of sources.
5:
6:     while S ≠ ∅ do
7:         x ← ∈ S
8:         S ← S \ { x }
9:         L ← L + { x }                    ▷ x is added at the end of L.
10:
11:        for (x, y) ∈ A do
12:            A ← A \ { (x, y) }
13:            if ∀_z ¬ ( (z, y) ∈ A ) then   ▷ y is a new source.
14:                S ← S ∪ { y }
15:            end if
16:        end for
17:
18:    end while
19:
20:    if A ≠ ∅ then
21:        G  has a cycle:  Report defeat
22:    else
23:        Report L
24:    end if
25:
26: end procedure
```

Exercise 2.14

Prove that any start–node may start a topological sort of a DAG.

A start–node x is selected as a first element in L. The vertex x is then removed from the graph and a topological sort is performed on the remaining graph. In the remaining graph, all arcs (x, y) are removed — since x is no longer a vertex of the graph — and the set of start–nodes of $G - x$ is determined. The process continues as long as the set of start–nodes is nonempty.

Exercise 2.15

Prove that upon completion of the algorithm described above — that is when $S = \varnothing$ — any remaining arc implies that G has a cycle.

Hint: The remaining digraph — if any — has no source.

Theorem 2.16. *There exists a linear–time algorithm* [19] *that computes a topological sort in a* DAG.

Proof. The graph is represented as a list of arcs. For each vertex the algorithm maintains a list of out–neighbors and a list of in–neighbors. These lists are updated in $O(1)$ time whenever an arc is removed.

To find the set of start–nodes, define a Boolean array

$$b : V \to \{ \text{TRUE}, \text{FALSE} \}$$

and initialize it as TRUE for all vertices. Then the algorithm passes through all arcs $(x, y) \in A$ and sets $b : (y) = \text{FALSE}$. The start–nodes are then the remaining vertices, ie, those x for which $b(x)$ remained TRUE. This part of the algorithm can be implemented so that it runs in $O(n + m)$ time.

At each pass of the loop in Kahn's algorithm, starting at Line 7, a vertex x is removed from S and added to L. Effectively, x is removed from the graph. The removal of arcs that leave x takes $O(1)$ time per arc. The fact that each arc is removed at most once proves the timebound.

This proves the theorem.   □

[19] By 'linear' we mean that the algorithm runs in $O(n + m)$ time.

Figure 2.4: The 8-wheel W_8 (The figure appears here for no particular reason.)

— By Theorem 2.16 — there exists a linear–time algorithm that solves the simple total ordering problem. Unfortunately — there is no method like that available for the total ordering problem.

Opatnrý shows that the problem to color a hypergraph of rank 3 can be reduced to the total ordering problem. We explore this issue in the next section.

2.4.1 Hypergraphs

Definition 2.17. A <u>hypergraph</u> H is a pair $(V, \mathcal{E})$ — where V is a finite nonempty set and $\mathcal{E} = E(H)$ is a set of nonempty subsets of V . [20]

The elements of $\mathcal{E}$ are called <u>hyperedges</u> .

[20] Hypergraphs are sets of subsets of a 'universal set' of vertices V .

Definition 2.18. The <u>rank</u> of a hypergraph H is the maximal cardinality of its hyperedges, ie,

$$\text{rank}(H) = \max\{|e| \mid e \in \mathcal{E}\},$$

where $\mathcal{E} = E(H)$ denotes the set of hyperedges of H .   (2.11)

The 2–coloring problem for hypergraphs $H = (V, \mathcal{E})$ is to find a partition of its vertices — say $\{A, B\}$ — such that each hyperedge has a nonempty intersection with A and B , ie,

$$\forall_{e \in \mathcal{E}} \quad e \cap A \neq \varnothing \quad \text{and} \quad e \cap B \neq \varnothing.   (2.12)$$

A hypergraph is called <u>k-uniform</u> if all its hyperedges have k vertices.

When H is 2–uniform then H is a graph and then the 2–coloring problem is easy to solve. [21]

[21] See exercise 1.14.

In the remainder of this section we show that the total ordering problem is at least as hard as the 2–coloring problem for hypergraphs of rank 3. By that we mean that if there exists a polynomial–time algorithm that solves the total ordering problem, then that algorithm can be adapted, so that it solves the 2–coloring problem for hypergraphs of rank 3 in polynomial time.

Exercise 2.16

An <u>orientation</u> of a graph gives each edge a direction. Show that any graph can be oriented into a DAG.

Exercise 2.17

Orient the graph in Figure 2.4 into a DAG. Implement Kahn's algorithm, and find a topological sort.

Remark 2.19. Cook showed that the 2–coloring problem for hypergraphs of rank 3 is NP–complete. There are some indications that show that there is — probably — no polynomial–time algorithm for any NP–complete problem. We will speak more about that in our chapter on complexity.

2.4.2 Problem Reductions

The method — to show that the total ordering problem is at least as hard as the 2–coloring problem — is called a <u>problem reduction</u>. We <u>reduce</u> the 2–coloring problem for hypergraphs of rank 3 to the total ordering problem and we show that this reduction takes polynomial time.

Assume that there exists a polynomial–time algorithm that solves the total ordering problem in time $O\left((n+|R|)^k\right)$, for some $k \in \mathbb{N}$. Here $n = |V|$, the cardinality of the universal set, and R is the set of betweenness constraints. We show that there also exists an algorithm that solves the 2-coloring problem for hypergraphs H of rank 3 in $O\left((n+m)^k\right)$ time. Here $n = |V(H)|$ and $m = |E(H)|$.

Proof. Let H be a hypergraph of rank 3 for which we wish to solve the 2–coloring problem. Notice that we may assume that there are no hyperedges of cardinality 1, otherwise there cannot exist a 2–coloring of H and we are done. Henceforth, we assume that every hyperedge is either a triple or a pair of elements of V.

Number the vertices and hyperedges of H:

$$V(H) = \{h_1, \cdots, h_n\} \quad \text{and}$$
$$E(H) = \{\, (a_i, b_i, c_i) \mid \{a_i, b_i, c_i\} \subseteq V \quad \text{and} \quad 1 \leqslant i \leqslant t\}$$
$$\bigcup \ \{(d_j, e_j) \mid \{d_j, e_j\} \subseteq V \quad \text{and} \quad 1 \leqslant j \leqslant p\}. \quad (2.13)$$

Construct a universal set and a set of betweenness constraints
as follows. For each triple $(a_i, b_i, c_i) \in E(H)$ introduce one
vertex, say y_i. Add one more vertex x. The universal set
V^* is

$$V^* = V(H) \cup \{x\} \cup \{y_i \mid (a_i, b_i, c_i) \in E(H)\}. \quad (2.14)$$

The set of betweenness constraints R is defined as follows.

For each triple $(a_i, b_i, c_i) \in E(H)$ the following two triples
are in R,

$$(a_i, y_i, b_i) \in R \quad \text{and} \quad (y_i, x, c_i) \in R. \quad (2.15)$$

For each pair $(d_i, e_i) \in E(H)$ the following triple is in R:

$$(d_i, x, e_i) \in R. \quad (2.16)$$

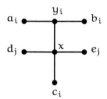

This completes the description of the betweenness relations and
the universal set.

We claim that H has a 2–coloring if and only if there is a
total ordering of V^* satisfying the betweenness constraints R.

Figure 2.5: The figure illustrates the betweenness constraints.

Assume that the hypergraph H has a 2–coloring of its vertices.
Let $\{A, B\}$ be a partition of V that contains an endpoint of
every hyperedge. We show that there is a linear ordering of
V^* satisfying R.

Exercise 2.18

Construct an injective map $f : V^* \to \mathbb{Q}$ as follows.

Show that the function f is injective.

$$f(x) = 0 \quad (2.17)$$

$$\forall_{h_\ell \in V} \quad f(h_\ell) = \begin{cases} \ell & \text{if } h_\ell \in A \\ -\ell & \text{if } h_\ell \in B, \end{cases} \quad (2.18)$$

$$\forall_{i \in [t]} \quad f(y_i) = \begin{cases} \min\{f(a_i), f(b_i)\} + \frac{1}{i+1} & \text{if } \text{sign}(f(a_i)) = \text{sign}(f(b_i)) \\ -\frac{\text{sign}(f(c_i))}{i+1} & \text{otherwise.} \end{cases} \quad (2.19)$$

$$\forall_{z \in \mathbb{Z}} \quad \text{sign}(z) = \begin{cases} 1 & \text{if } z > 0 \\ -1 & \text{if } z < 0. \end{cases}$$

Notice that each vertex of H is mapped to i or to $-i$, for
some natural number $i \in \mathbb{N}$. It is positive when it is in

A and negative otherwise. Since the endpoints of each hyper-edge $(c_i, d_i) \in E(H)$ have opposite colors each betweenness constraint $(c_i, x, d_i) \in R$ is satisfied.

Consider $(a_i, b_i, c_i) \in E(H)$. We check if the betweenness constraint (a_i, y_i, b_i) is satisfied.

If a_i and b_i are both in the same sets of the partition, ie, if

$$\mathrm{sign}(f(a_i)) = \mathrm{sign}(f(b_i))$$

then — since $f(a_i) \neq f(b_i)$ [22] — y_i is mapped to the smallest of $f(a_i)$ and $f(b_i)$ plus $1/(i+1)$, and so $f(y_i)$ lies between $f(a_i)$ and $f(b_i)$.

[22] $f(a_i) \neq f(b_i)$ because $a_i \neq b_i$.

If $\mathrm{sign}(f(a_i)) \neq \mathrm{sign}(f(b_i))$, then

$$f(y_i) \in (-1, 0) \cup (0, 1)$$

— and so — y_i lies between $f(a_i)$ and $f(b_i)$ which are integers of opposite sign.

Exercise 2.19

Check that also all betweenness constraints (y_i, x, c_i) are satisfied.

This completes the proof that V^* has a total order, namely, since $\mathbb{Q}$ is a total order, the function f defines a total ordering of V^*.

Assume now that there is a total order of V^* — say $\leqslant$. [23] Define a partition $\{A, B\}$ of $V(H)$ by

[23] We're not done yet!

$$A = \{h \mid h \in V(H) \text{ and } h > x\} \quad \text{and} \quad B = V \setminus A. \quad (2.20)$$

By the betweenness constraints each hyperedge

$$(d_i, e_i) \in E(H)$$

has x between the two endpoints — and so — only one of the two endpoints is in A.

Similarly, there cannot be a hyperedge

$$(a_i, b_i, c_i),$$

with all three elements $< x$ or all three $> x$.

Exercise 2.20

Check all four cases, that no hyperedge $\{a, b, c\} \in E(H)$ is monochromatic.

— We are done — we have shown that given a hypergraph of rank 3 we can construct an instance for the total ordering problem in linear time. The universal set V^* has a total ordering if and only if the hypergraph has a 2–coloring. □

Exercise 2.21

Implement Opatrný's reduction — Start with the following hypergraph (V, E):

$$V = [n] \tag{2.21}$$
$$E = \{\{i, i+1, i+2\} \mid 1 \leqslant i \leqslant n-2\}. \tag{2.22}$$

Make a list of the betweenness constraints. What is the universal set and what are the total orders that satisfy all the betweenness constraints?

2.5 NP – *Completeness*

Let P be some problem that we wish to solve. — For simplicity — assume that P is a 'yes-or-no' question — or —

a <u>decision problem</u>. [24]

We would like to have a fast algorithm — eg, polynomial — to solve P. The question whether we can do that concerns the — complexity — of the problem.

[24] Usually, the decision variant of a problem is sufficient to solve it. For example, to compute $\omega(G)$ in polynomial time, it would be sufficient to have a polynomial-time algorithm that checks if $\omega(G) \geqslant k$, for each $k \in [n]$.

To acquire some information about the complexity of a
decision problem P, it is useful to consider the alternative
problem P*, which is the problem P equipped with an oracle [25]
that gives you the answer to P. Then the remaining question
that you need to solve is whether the oracle has given you
the correct answer.

[25] eg, your teacher

Definition 2.20. The class of problems NP is the class of
decision problems P for which an answer — supplied by an
oracle — can be tested in polynomial time.

Definition 2.21. A decision problem P is NP – complete if
it is in NP and every other problem in NP reduces to it
in polynomial time.

Example 2.22. For example, in Section 2.4.2 on Page 35 we
reduced the 2 – coloring problem for hypergraphs of rank 3, to
the total ordering problem. — Given the fact that — the
2 – coloring problem is NP – complete – we have shown that the
total ordering problem is NP – complete as well.

2.5.1   Equivalence covers of splitgraphs

Definition 2.23. A graph is an equivalence graph if it is P_3-free
— that is — if it is a disjoint union of cliques. An equivalence cover
of a graph G is a set of equivalence subgraphs that covers E(G).

The minimal number of equivalence graphs in a cover of G is
denoted as q(G).

Definition 2.24. A graph is a splitgraph if there is a partition of
its vertices into a clique and an independent set.

Exercise 2.22

Let G be a splitgraph and let $\{K, S\}$ be a partition of $V(G)$ such that K induces a clique and S induces an independent set. For a vertex $x \in K$ define

$$\delta(x) = |N(x) \cap S| \qquad \text{and let} \qquad D = \max\{\delta(x) \mid x \in K\}.$$

1. Show that $q(G) \geqslant \Delta$: Choose a vertex $x \in K$ that has D neighbors in S. The star induced by x and its neighbors in S has equivalence number D.

2. Show that $q(G) \leqslant D + 1$: To see that enumerate the vertices of S — say $y_1 \cdots y_t$. For each $x \in K$ order its neighbors in S in some arbitrary order. For $i = 1, \cdots, D$, define the equivalence graph with cliques

$$W_{i,j} = \{y_j\} \cup \{x \in K \mid \text{the } i^{\text{th}} \text{ neighbor of } x \text{ is } y_j\}.$$

Define one more equivalence graph that consists of one clique; namely K.

> In this section we show that computing the equivalence cover number of splitgraphs is NP-complete.

Chromatic index

The chromatic index of a graph is the minimal number that is needed to color the edges of a graph such that no two edges with a nonempty intersection have the same color. The chromatic index is denoted as $\chi'(G)$. By Vizing's theorem the chromatic index of a graph is either Δ or $\Delta + 1$, where Δ is the maximal degree of a vertex in the graph.

Holyer proved the following theorem. [26]

Theorem 2.25. *It is NP-complete to decide whether the chromatic index of a cubic graph is 3 or 4.*

[26] I. Holyer, *The NP-completeness of edge-coloring*, SIAM J. Comput. **10** (1981), pp. 718–720.

A graph is cubic if every vertex has degree 3. For example, the Petersen graph is cubic.

Exercise 2.23

Assume that G is triangle - free. Show that

$$q(G) \quad = \quad \chi'(G).$$

Exercise 2.24

Let G be a cubic graph. Construct a graph G' as follows. Introduce a new vertex x_e for every edge $e \in E(G)$. Make a new vertex x_e adjacent to the two endpoints of e.

Show that

$$\chi'(G) = 3 \quad \Leftrightarrow \quad q(G') = 3.$$

Corollary 2.26. *It is* NP-*complete to decide whether a graph without* K_4 *and with maximal degree at most 6 has equivalence cover number 3 or 4.*

LET G BE A CUBIC GRAPH. CONSTRUCT A <u>SPLITGRAPH</u> G^* AS FOLLOWS.

S1. The splitgraph G^* has a clique $K = V(G)$.

S2. For each $e \in E(G)$ the independent set S of G^* contains two vertices x_e and y_e which are both adjacent to the endpoints of e in K.

S3. For each nonedge $f \in E(\bar{G})$ the independent set S of G^* contains one vertex z_f which is adjacent to the endpoints of f in K.

This completes the description of the splitgraph G^*.   Let's hope it works!

Exercise 2.25

Let G be a cubic graph and let G^* be the graph constructed as above.

$$\chi'(G) = 3 \quad \Leftrightarrow \quad q(G^*) = n + 2$$

where $n = |V(G)|$.

This proves the following theorem.

A. Blokhuis and T. Kloks, *On the equivalence covering number of splitgraphs*, Information Processing Letters **54** (1995), pp. 301–304.

Theorem 2.27. *It is* NP-*complete to decide whether the equivalence cover number of a splitgraph is* D *or* D + 1. *This remains* NP-*complete when the class is restricted so that all vertices in the independent sets of the splitgraphs have degree 2.*

2.6 Lovász Local Lemma

TO SHOW THE EXISTENCE of combinatorial objects the Lovász Local Lemma can be of great use.

TO START consider the 2-coloring problem of a hypergraph $\mathcal{H}$: we wish to color the vertices with two colors such that no hyperedge is monochromatic. Assume that $\mathcal{H}$ is k-uniform — that is — every hyperedge of $\mathcal{H}$ has k vertices.

When $\mathcal{H}$ has less than 2^{k-1} edges then $\mathcal{H}$ is 2-colorable.

To see that, color the vertices of $\mathcal{H}$ independently with probability 1/2 red or black. A BAD EVENT is an hyperedge that is colored monochromatic.

The hypergraph is k-uniform — so — the probability that a bad event occurs is at most 2^{1-k} (either all vertices of the hyperedge are colored black or all vertices are colored red). The probability that <u>some</u> bad event happens is at most their sum and this is less than one (by the assumption on the number of hyperedges). The conclusion is that there is a 2-coloring of $\mathcal{H}$ in which no bad event occurs.

This result is not so great: for a graph it simply says that it is bipartite whenever it has less than two edges (in which case there can be no cycle). It is easy to do better; when each edge intersects at most one other edge then the graph is bipartite also. A generalization of this case is captured by Lovász' local lemma.

Assume that every hyperedge of a k-uniform hypergraph $\mathcal{H}$ intersects at most d other hyperedges. Lovász' local lemma allows us to conclude that $\mathcal{H}$ is 2-colorable whenever

$$e \cdot (d+1) \quad \leqslant \quad 2^{k-1}$$

(where $e = 2.718\cdots$ is the basis of the natural logarithm).

IN A PROBABILITY SPACE consider a finite set of mutually independent random variables. Let

$$A_1 \quad \cdots \quad A_n$$

be a collection of <u>events</u>. An event is determined by the values of a subset of the variables in the outcome of an experiment. We write $P(A_i)$ for the probability that an event A_i occurs.

The A_i are the (bad) events that we wish to avoid. (In the example above the bad events are monochromatic hyperedges that turn up in the outcome of an experiment which colors the vertices of the hypergraph.)

We can avoid all bad events if we prove

$$P(\cap \bar{A}_i) \quad > \quad 0.$$

When the events are independent then their complements $\bar{A}_i$ are also independent. In that case

$$P(\cap \bar{A}_i) \quad = \quad \prod P(\bar{A}_i) \quad > \quad 0$$

— that is — there exists a way to assign values to the variables such that no bad event happens (unless some A_i surely happens). — On the other hand — it is clearly impossible to avoid all events when some subset of the $\bar{A}_j$s implies some (other) event A_i. Therefore we need some upperbound for the conditional probabilities

$$P(A_i \mid \bigcap_{j \in J} \bar{A}_j)$$

for any set $J \subseteq [n] \setminus i$.

The local lemma deals with the case where the events are 'almost' independent. To formalize this we make use of a <u>dependency graph</u>.

The dependency graph has a vertex for each subset of variables that determines an event. Two vertices are adjacent in the graph when the intersection of the two subsets is nonempty.

Definition 2.28. An event A is independent of a collection of events $B_1 \cdots B_k$ if for all $J \subseteq [k]$ and $J \neq \varnothing$

$$P(A \bigcap_{j \in J} B_j) \;=\; P(A) \;\times\; P(\bigcap_{j \in J} B_j).$$

Definition 2.29. Let $A_1 \cdots A_n$ be events in a probability space. A graph $D = (V, E)$ with $V = [n]$ is a dependency graph if each event A_i is independent of the collection of events

$$\{A_j \;\mid\; \{i, j\} \;\notin\; E\}.$$

Spencer formulated and proved Lovász' local lemma (originally proved by Lovász and Erdős) as follows.

Lemma 2.30. *Let $A_1 \cdots A_n$ be events with a dependency graph. Let $0 \leqslant x_i < 1$ be real numbers assigned to the events such that*

$$P(A_i) \;\leqslant\; x_i \cdot \prod_{\{i,j\} \in E} (1 - x_j).$$

Then

$$\prod P(\cap \bar{A}_i) \;\geqslant\; \prod (1 - x_i) \;>\; 0.$$

Proof. We first show that for any $J \subseteq [n] \setminus i$

$$P(A_i \mid \cap_{j \in J} \bar{A}_j) \;\leqslant\; x_i.$$

This is true when $J = \varnothing$, since

$$P(A_i) \;\leqslant\; x_i \cdot \prod_{\{i,j\} \in E} (1 - x_j) \;\leqslant\; x_i.$$

We proceed by induction on $|J|$. Let

$$J_1 \;=\; \{j \in J \mid (i, j) \in E.\} \quad \text{and let} \quad J_2 = J \setminus J_1$$

We may assume that $J_1 \neq \varnothing$ otherwise the claim is clearly true. We can write

$$P(A_i \mid \cap_{j \in J} \bar{A}_j) \;=\; \frac{P(A_i \cap_{j \in J_1} \bar{A}_j \mid \cap_{j \in J_2} \bar{A}_j)}{P(\cap_{j \in J_1} \bar{A}_j \mid \cap_{j \in J_2} \bar{A}_j)}. \qquad (2.23)$$

The event A_i is independent of the set of events $\{A_j \mid j \in J_2\}$. We use that to find an upperbound for the numerator in (2.23).

$$P(A_i \cap_{j \in J_1} \bar{A}_j \mid \cap_{j \in J_2} \bar{A}_j) \leqslant P(A_i \mid \cap_{j \in J_2} \bar{A}_j)$$
$$= P(A_i) \leqslant x_i \cdot \prod_{\{i,j\} \in E} (1 - x_j).$$

To find a lowerbound for the denominator in (2.23), write

$$J_1 = \{j_1, \cdots, j_r\}.$$

Using induction we obtain

$$P(\bar{A}_{j_1} \cap \cdots \cap \bar{A}_{j_r} \mid \cap_{j \in J_2} \bar{A}_j) =$$
$$P(\bar{A}_{j_1} \mid \cap_{j \in J_2} \bar{A}_j) \times P(\bar{A}_{j_2} \mid \bar{A}_{j_1} \cap_{j \in J_2} \bar{A}_j) \times$$
$$\cdots \times P(\bar{A}_{j_r} \mid \bar{A}_{j_1} \cap \cdots \cap \bar{A}_{j_{r-1}} \cap_{j \in J_2} \bar{A}_j)$$
$$\geqslant \prod_{\{i,j\} \in E} (1 - x_j).$$

This proves the claim.

The following observation completes the proof of the lemma.

$$P(\cap \bar{A}_i) = P(\bar{A}_1) \times P(\bar{A}_2 \mid \bar{A}_1) \times \cdots$$
$$\times P(\bar{A}_n \mid \bar{A}_1 \cap \cdots \cap \bar{A}_{n-1}) \geqslant \prod_{i=1}^{n} (1 - x_i) > 0.$$

$\square$

Remark 2.31. Assume that all degrees in a dependency graph are at most d. Assume that the probability of any bad event satisfies

$$P(A_i) \leqslant p$$

We claim that if $e \cdot p \cdot (d+1) \leqslant 1$ then $P(\cap \bar{A}_i) > 0$. To see that set $x_i = 1/d+1$. Since the degree of any vertex is at most d, we have

$$x_i \cdot \prod_{\{i,j\} \in E} (1 - x_j) \geqslant \frac{1}{d+1} \cdot (1 - \frac{1}{d+1})^d \geqslant$$
$$\frac{1}{e \cdot (d+1)} \geqslant p.$$

So the local lemma applies.

Exercise 2.26

Prove the claim we made at the start of this section: Let $\mathcal{H}$ be a k-uniform hypergraph and assume that every hyperedge intersects at most d other hyperedges. When $e(d+1) \leqslant 2^{k-1}$ then $\mathcal{H}$ is 2-colorable.

Remark 2.32. In 2009 Moser and Tardos presented a constructive proof of the Lovász local lemma — that is — they present an algorithm that finds a good object efficiently.

Remark 2.33. Thomassen shows that every hypergraph which is k-regular and k-uniform (that is; every hyperedge has k vertices and every vertex is in k hyperedges) is 2-colorable <u>provided</u> $k \geqslant 4$. The Fano plane shows that not every 3-regular, 3-uniform hypergraph is 2-colorable. The 2-regular graphs that are not 2-colorable are — of course — the odd cycles.

2.6.1   Bounds on dominating sets

The following problem is an example of a problem which can be tackled using the Lovász Local Lemma.

Definition 2.34. Let $a, b \in \mathbb{N}$. A set S of vertices in a graph is an (a, b)-dominating set if every vertex of S is adjacent to at least a vertices in S and every vertex outside S is adjacent to at least b vertices in S.

For a graph G with all degrees at least a let $\gamma_{a,b}(G)$ be the smallest number of elements in an (a, b)-dominating set in G.

Lemma 2.35. *Let* $\frac{1}{2} \leqslant \alpha < 1$. *There exists a number* $R \geqslant 0$ *such that for all* $r > R$ *any* r-*regular graph satisfies*

$$\gamma_{a,b} \quad \leqslant \quad \alpha \cdot n.$$

Proof. LET $N \in \mathbb{N}$ and color (independently) the vertices of an r-regular graph G with N colors. One of the colors appears at least n/N times among the vertices of G. Say red is such a color and write $V \setminus \text{red}$ for the vertices that are not red.

We CLAIM that

$$V \setminus \text{red}$$

is with positive probability an (a, b)-dominating set.

For a vertex x define a bad event A_x as a coloring of $N[x]$ such that either one of the following holds.

1. x is red and x has less than b neighbors in $V \setminus \text{red}$

2. x is not red and x has less than a neighbors in $V \setminus \text{red}$

The following formula expresses the probability that a bad event A_x occurs.

$$P(A_x) = \frac{1}{N^{r+1}} \cdot \left(\sum_{i=0}^{b-1} \binom{r}{i} (N-1)^i + (N-1) \cdot \sum_{i=0}^{a-1} \binom{r}{i} (N-1)^i \right). \quad (2.24)$$

Notice that each event A_x is dependent of at most r^2 other events; namely the events A_y for vertices y at distance at most two from x. — So — by Remark 2.31 (or by Lovász local lemma with all variables $x_i = 1/r^2$) we are done when we show that

$$e \cdot P \cdot r^2 \leqslant 1,$$

where we write $P = P(A_x)$ for the probability of a bad event as in Formula (2.24).

For any given number $N \geqslant 2$ since a and b are fixed numbers in the formula (2.24) the numerator is a polynomial in r and the denominator is an exponential function in r. — So — there exists a number R such that $e \cdot P \cdot r^2 < 1$ whenever $r > R$.

This proves the lemma.   □

It is easy to cover the case where G is a graph with minimal degree δ and maximal degree Δ: Use δ to adjust (2.24) and use Δ^2 to bound the degree in the dependency graph.

An (a,b)-dominating set that meets the requirements can be constructed efficiently by the algorithm of Moser and Tardos.

2.6.2   The Moser & Tardos algorithm

Moser and Tardos developed (in 2009) a constructive proof of Lovász' local lemma. — To be more precise — let $\mathcal{A}$ be a collection of events with a dependency graph D.

To ease the notations we identify an event A with the set of variables whose outcome determines A.[27]

For an event $A \in \mathcal{A}$ we write $N(A) \subseteq \mathcal{A} \setminus A$ for its neighbors in the dependency graph D and we write $\bar{N}(A) = N(A) \cup \{A\}$ for its closed neighborhood.[28] Let $x : \mathcal{A} \to (0,1)$ be a function which satisfies Lovász' condition

$$\forall_{A \in \mathcal{A}} \quad P(A) \quad \leqslant \quad x(A) \quad \times \prod_{B \in N(A)} (1 - x(B)). \qquad (2.25)$$

MOSER AND TARDOS show that a very simple randomized algorithm finds an assignment of the variables which avoids all events $A \in \mathcal{A}$. — Furthermore — the expected number of <u>resampling steps</u> used by this algorithm is bounded by

$$\sum_{A \in \mathcal{A}} \frac{x(A)}{1 - x(A)}.$$

In this section we recapitulate Moser and Tardos' result. In their paper they show furthermore that when the dependency graph has bounded degree and the function x satisfies a slightly stronger condition than (2.25) then there is an 'efficient' deterministic algorithm that finds an assignment of the variables which avoids all events of $\mathcal{A}$. (We refer to their paper for a precise description of this and other results.)

We say that an experiment <u>violates</u> an event when the event happens.

[27] In their paper Moser and Tardos introduce the notation $\mathsf{vbl}(A)$ for this set of variables.

[28] In the dependency graph there is an edge between two events A and B when $A \cap B \neq \varnothing$. So an event A is independent of the collection $\mathcal{A} \setminus (\{A\} \cup \Gamma(A))$. (See Definiton 2.28 on Page 44.)

THE RANDOMIZED ALGORITHM to find an assignment of the variables which avoids all the events of $\mathcal{A}$ is the following.

---

1. Start with a random asignment of all variables

2. If some event of $\mathcal{A}$ occurs then pick one arbitrarily and find a new random assignment of the variables that it contains

3. Repeat this resampling step until no more bad event occurs.

BELOW we prove the following theorem.

Theorem 2.36. *Let $\mathcal{P}$ be a finite set of mutually independent random variables in a probability space. Let $\mathcal{A}$ be a finite set of events and let x be a function which satisfies the condition of Lovász local lemma (2.25). The algorithm described above resamples an event A an expected number of times at most $x(A)/{1-x(A)}$ — that is — the expected total number of resampling steps is at most*

$$\sum_{A \in \mathcal{A}} \frac{x(A)}{1 - x(A)}.$$

2.6.3 Logs and witness trees

TO PROVE THEOREM 2.36 we show that the algorithm is equivalent to 'checking' sequences of 'witness trees' for the occurrence of bad events.

The execution of the algorithm above produces a log — that is — a sequence of events

$$C : \mathbb{N} \quad \rightarrow \quad \mathcal{A},$$

which are chosen for resampling during the execution. This is a partial function when the algorithm terminates. We assume that there is a fixed (randomized) procedure which selects the bad event for resampling. (This makes the log a random variable.)

Definition 2.37. A <u>witness tree</u> is a finite rooted tree T with a labeling

$$[\cdot] : V(T) \quad \to \quad \mathcal{A}$$

such that for each node a its children are labeled by elements of $\bar{N}([a])$.

A witness tree is called <u>proper</u> if all children of a node in a witness tree have different labels.

GIVEN A LOG C we identify a witness tree with each element $C(t)$ as follows. Start with a tree T^t which consists of a single root node with label $C(t)$. For $i = t - 1, \cdots 1$ distinguish two cases to construct the tree T^i.

- If there is a vertex a in the tree T^{i+1} with $C(i) \in \bar{N}([a])$ then choose a such that it is furthest from the root. Attach a new child to a and label it as $C(i)$.

- If $\bar{N}(C(i)) \cap V(T^{i+1}) = \varnothing$ then let $T^i = T^{i+1}$.

The witness tree $\tau(t)$ of the resampling step t in C is defined as

$$\tau(t) \quad = \quad T^1.$$

Definition 2.38. A witness tree T <u>appears in C</u> if there exists some $t \in \mathbb{N}$ such that $T = \tau(t)$.

Lemma 2.39. *A witness tree which appears in a log is proper.*

Proof. By definition of the algorithm that produces the witness tree; no two elements of $\mathcal{A}$ that are the same or intersect can appear at the same depth in a witness tree.   □

WE WANT TO SHOW that the probability that a witness tree T appears in C is at most the probability that it passes a certain test.

Definition 2.40. Let T be a witness tree. A <u>T-check</u> visits the nodes of T in <u>reversed BFS-order</u>, it takes a random evaluation of elements of each node [a] in T that it visits and checks if the event [a] is violated. The witness tree T <u>passes the check</u> if all events were violated when checked.

Lemma 2.41. *The probability that a witness tree* I *passes its check is*

$$\prod_{a \in V(T)} P([a]).$$

Proof. The random evaluation of the variables in each node is independent of earlier evaluations. □

Lemma 2.42. *Let* T *be a fixed witness tree and let* C *be the random log produced by the algorithm. The probability that* T *appears in the log is at most*

$$\prod_{a \in V(T)} P([a]).$$

Proof. Assume that a random generator produces an infinite sequence of independent random evaluations for each variable P

$$P^1 \quad P^2 \quad \dots$$

Whenever the algorithm of Moser and Tardos (or a T-check) calls for a new random sample of P the generator presents the next element in the list.

Assume that a witness tree T appears in the log C — say $T = \tau(t)$ for some $t \in \mathbb{N}$. (So the root of T is labeled as $C(t)$). We need to prove that T passes the T-check.

Let $a \in V(T)$ and let $P \in [a]$. Let $S(P)$ be the set of nodes $w \in V(T)$ that are at depth greater than v in T and for which $P \in [w]$. When the T-check visits v the random evaluation of P produces $P^{|S(P)|}$. That is so because the check visits the nodes in order of decreasing depth and no two nodes at the same depth can contain the same element P (since they are disjoint).

Notice that the randomized algorithm of Moser and Tardos re-samples the variable P exactly in the same order (by definition of the occurrence of T in the log). Since [a] is violated in the Moser & Tardos algorithm it is also violated in the T-check.

This proves the lemma.   □

Let N_A be the number of times that an event $A \in \mathcal{A}$ appears in the log — that is — N_A is the number of times that A is resampled during the execution of the Moser & Tardos algorithm. Then N_A is equal to the number of different witness trees that appear in C and that have their root labeled A. To see that let t_i be such that $C(t_i) = A$ for the ith time. The witness tree $\tau(t_i)$ contains exactly i copies of A — so clearly — $\tau(t_i) \neq \tau(t_j)$ whenever $i \neq j$.

It follows that we can bound the expectation of N_A by summing the bounds on the probabilities of occurrences of witness trees. We do that in the next section.

2.6.4 A Galton - Watson branching process

Wassup ?

In this section we describe and analyze a process that generates proper witness trees with a fixed root $A \in \mathcal{A}$.

Let $x(\cdot)$ be a function satisfying the Lovász' condition 2.25 on Page 48. A process to generate a witness tree is the following.

---

1. In the first round the process generates a tree with a single node labeled A

2. For each subsequent round choose (independently) an element a of the previous round and (also independently) choose an element $B \in \bar{N}([a])$

3. With probability $x(B)$ add a child node at the node a and give it the label B. With probability $1 - x(B)$: skip.

This process ends when no new vertices are created in some round.

Define

$$x'(B) \quad = \quad x(B) \quad \cdot \quad \prod_{C \in N(B)} (1 - x(C)).$$

Lemma 2.43. *Let* T *be a proper witness tree with its root labeled* A. *The probability that the Galton-Watson process described above produces exactly* T *is*

$$P_T \quad = \quad \frac{1 - x(A)}{x(A)} \quad \cdot \quad \prod_{a \in V(T)} x'([a]).$$

Proof. For $a \in V(T)$ let W_a be the set of elements in $\bar{N}([a])$ that do not appear as the label of a child of a. The probability that the Galton-Watson process produces T is

$$P_T \quad = \quad \frac{1}{x(A)} \quad \cdot \quad \prod_{a \in V(T)} x([a]) \cdot \prod_{u \in W_a} (1 - x([u])).$$

(The leading factor appears because the root is always there.)

We get rid of the W_a's as follows

$$P_T \quad = \quad \frac{1 - x(A)}{x(A)} \quad \cdot \quad \prod_{a \in V(T)} \frac{x([a])}{1 - x([a])} \cdot \prod_{u \in \bar{N}([a])} (1 - x([u])).$$

We can replace inclusive neighborhoods by exclusive ones

$$P_T \quad = \quad \frac{1 - x(A)}{x(A)} \quad \cdot \quad \prod_{a \in V(T)} x([a]) \cdot \prod_{u \in N([a])} (1 - x([u])) =$$
$$\frac{1 - x(A)}{x(A)} \quad \cdot \quad \prod_{a \in V(T)} x'([a]).$$

This proves the lemma. $\qquad\qquad\qquad\qquad\qquad\square$

We are now ready to complete the proof of Theorem 2.36 (which is on Page 49).

Proof. Let $\mathcal{T}_A$ be the set of all proper witness trees that have root A. We find for the expected number of times that an event A appears in the log C:

$$\mathbb{E}(N_A) \;=\; \sum_{T \in \mathcal{T}_A} P(T \text{ appears in } C) \;\leqslant\;$$

$$\sum_{T \in \mathcal{T}_A} \prod_{a \in V(T)} P([a]) \;\leqslant\; \sum_{T \in \mathcal{T}_A} \prod_{a \in V(T)} x'([a]) \;=\;$$

$$\frac{x(A)}{1-x(A)} \cdot \sum_{T \in \mathcal{T}_A} P_T \;\leqslant\; \frac{x(A)}{1-x(A)}.$$

The first inequality follows from Lemma 2.42. The second inequality follows from the assumptions in Theorem 2.36. The third inequality follows from Lemma 2.43. The last inequality holds because the Galton-Watson process produces exactly one tree at a time.

This proves the theorem. □

2.7 Szemerédi's Regularity Lemma

The regularity lemma plays an important role in extremal combinatorics. [29]

Let G be a graph and let X and Y be disjoint sets of vertices. Write $e(X,Y)$ for the number of edges that intersect both X and Y. The <u>density</u> of $\{X,Y\}$ is defined as $d(X,Y) = e(X,Y)/|X|\cdot|Y|$.

Definition 2.44. Let X and Y be disjoint sets. The pair $\{X,Y\}$ is <u>ϵ-regular</u> if for all $X' \subseteq X$ and $Y' \subseteq Y$

$$|X'|/|X| \;\geqslant\; \epsilon \quad \text{and} \quad |Y'|/|Y| \;\geqslant\; \epsilon \;\Rightarrow$$
$$|\, d(X',Y') - d(X,Y)\,| \;\leqslant\; \epsilon.$$

Exercise 2.27

Show that $\{X,Y\}$ is ϵ-regular when $d(X,Y) \leqslant \epsilon^3$.

[29] Extremal graph theory studies maximal or minimal graphs satisfying a certain property.

Definition 2.45. A partition $\{V_0, \cdots, V_k\}$ of the vertices of a graph is <u>equitable</u> if

$$|V_i| = |V_j| \quad \text{for all } 1 \leqslant i < j \leqslant k.$$

The set V_0 may be empty; it is called the exceptional class of the partition.

Definition 2.46. An equitable partition $\{V_0, \cdots, V_k\}$ is <u>ϵ-regular</u> if both of the following conditions hold.

(a) $|V_0| \leqslant \epsilon \cdot n$;

(b) all — except at most $\epsilon \cdot k^2$ of the pairs $\{V_i, V_j\}$ $(1 \leqslant i < j \leqslant k)$ —are ϵ-regular.

In this chapter we prove Szemerédi's regularity lemma:

THE REGULARITY LEMMA

Lemma 2.47. *Let $\epsilon \in \mathbb{R}$ and $t \in \mathbb{N}$. There exist $N, T \in \mathbb{N}$ such that any graph with at least N vertices has an ϵ-regular partition $\{V_0, \cdots, V_k\}$ with $t \leqslant k \leqslant T$.*

We will assume — throughout this chapter — that

$$0 \;<\; \epsilon \;\leqslant\; \frac{1}{2}.$$

This is not an important restriction since any ϵ'-regular partition is ϵ-regular for $\epsilon \geqslant \epsilon'$.

Exercise 2.28

Assume G has at most $\epsilon^4 \cdot n^2$ edges. Show that any equitable partition with $|V_0| \leqslant \epsilon \cdot n$ is ϵ-regular.

A partition π' is a <u>refinement</u> of another partition π if every class of π is the union of some classes of π'.

Definition 2.48. Let $\pi = \{V_1, \cdots, V_k\}$ be a partition of the vertices of a graph. The <u>index</u> of π is

$$\mathrm{index}(\pi) \quad = \quad \sum_{1 \leqslant i < j \leqslant k} \frac{|V_i| \cdot |V_j|}{n^2} \cdot d^2(V_i, V_j).$$

IMPORTANT MODIFICATION:

When $\pi = \{V_0, \cdots, V_k\}$ is a partition with an exceptional class V_0 then we define $\mathrm{index}(\pi)$ as the index of the refined partition where each element of V_0 forms a class by itself (so $\mathrm{index}(\pi)$ is the index of a partition with $|V_0| + k$ classes).

Exercise 2.29

Show that

$$0 \quad \leqslant \quad \mathrm{index}(\pi) \quad \leqslant \quad \frac{1}{2}.$$

IN THE SEARCH FOR AN ϵ-REGULAR PARTITION we start with an arbitrary equitable partition of the vertices $\{V_0, \cdots, V_t\}$ with

$$|V_0| \quad = \quad t - 1.$$

We refine this partition until it satisfies the conditions. To prove that the number of classes in the final partition is <u>independent</u> of n we use the fact that the index increases whilst it is bounded from above by $1/2$.

We first show that the index does not decrease in any refinement. — Below (in Lemma 2.51) — we show that the increase is substantial when there are 'irregular pairs' — that is — pairs $\{V_i, V_j\}$ with subsets V_i' and V_j' that satisfy

$$|V_i'| \quad \geqslant \quad \epsilon \cdot |V_i| \quad \text{and} \quad |V_j'| \quad \geqslant \quad \epsilon \cdot |V_j| \quad \text{and}$$
$$|d(V_i', V_j') - d(V_i, V_j)| \quad > \quad \epsilon. \quad (2.26)$$

Lemma 2.49. *If π' refines π then*

$$\text{index}(\pi') \quad \geqslant \quad \text{index}(\pi).$$

Proof. Consider a bipartition $\{X, Y\}$. Let $\{X_1, X_2\}$ be a partition of X. Then

$$e(X, Y) \quad = \quad e(X_1, Y) \quad + \quad e(X_2, Y).$$

Rewrite this as

$$|X||Y| \cdot d(X, Y) \quad = \quad |X_1||Y| \cdot d(X_1, Y) \quad + \quad |X_2||Y| \cdot d(X_2, Y).$$

By the Cauchy-Schwartz inequality we obtain

$$d^2(X, Y) \quad \leqslant \quad \frac{|X_1|}{|X|} \cdot d^2(X_1, Y) \quad + \quad \frac{|X_2|}{|X|} \cdot d^2(X_2, Y). \qquad (2.27)$$

Any refinement is obtained by repeated application of bipartitions. This proves the lemma. $\qquad\qquad \square$

Exercise 2.30

The Cauchy-Schwartz inequality says that for any real numbers a_i and b_i,

$$(a_1 b_1 + \cdots + a_n b_n)^2 \quad \leqslant \quad (a_1^2 + \cdots + a_n^2) \cdot (b_1^2 + \cdots + b_n^2).$$

Derive (2.27) (by suitable choice of a_i and b_i).

Lemma 2.50. *Let $\{X, Y\}$ be an irregular bipartition: let $\{X_1, X_2\}$ and $\{Y_1, Y_2\}$ be partitions of X and Y such that*

(a) $|X_1| \geqslant \epsilon \cdot |X|$ *and* $|Y_1| \geqslant \epsilon \cdot |Y|$;

(b) $|d(X_1, Y_1) - d(X, Y)| \quad \geqslant \quad \epsilon,$

then

$$\sum \frac{|X_i||Y_j|}{|X||Y|} \cdot d^2(X_i, Y_j) \quad \geqslant \quad d^2(X, Y) + \epsilon^4.$$

Proof. Notice that

$$|X\|Y| \cdot d(X, Y) \quad = \quad \sum |X_i\|Y_j| \cdot d(X_i, Y_j).$$

We have

$$\epsilon^4 \quad \leqslant \quad \sum \frac{|X_i\|Y_j|}{|X\|Y|} \cdot (d(X_i, Y_j) - d(X, Y))^2 \quad =$$

$$\sum \frac{|X_i\|Y_j|}{|X\|Y|} \cdot d^2(X_i, Y_j) \quad - \quad 2\,d(X, Y) \sum \frac{|X_i\|Y_j|}{|X\|Y|} \cdot d(X_i, Y_j) \quad +$$

$$d^2(X, Y) \quad =$$

$$\sum \frac{|X_i\|Y_j|}{|X\|Y|} \cdot d^2(X_i, Y_j) \quad - \quad d^2(X, Y).$$

This proves the lemma. $\qquad\qquad\qquad\qquad\qquad\qquad\qquad\qquad\square$

NOTATION: For a partition $\pi = \{V_0, \cdots, V_k\}$ with an exceptional class V_0 let $|\pi| = k$.

Lemma 2.51. *Let* $\pi = \{V_0, \cdots, V_k\}$ *be an equitable partition with an exceptional class* V_0 *that satisfies* $|V_0| \leqslant \epsilon \cdot n$. *Assume that there are more than* $\epsilon \cdot k^2$ *irregular pairs. There exists a refinement* π' *of* π *that satisfies*

$$\mathrm{index}(\pi') \quad > \quad \mathrm{index}(\pi) + \frac{1}{4} \cdot \epsilon^5 \quad \text{and} \quad |\pi'| \quad \leqslant \quad k \cdot 2^k.$$

Proof. When sets V_i and V_j form an irregular pair $\{V_i, V_j\}$ then there are partitions $\{V_i^1, V_i^2\}$ and $\{V_j^1, V_j^2\}$ such that

$$|V_i^1| \geqslant \epsilon|V_i|, \quad |V_j^1| \geqslant \epsilon|V_j|, \qquad |d(V_i^1, V_j^1) - d(V_i, V_j)| \quad > \quad \epsilon.$$

By Lemma 2.50 this implies

$$\sum_{k,\ell \in \{1,2\}} \frac{|V_i^k\|V_j^\ell|}{n^2} \cdot d^2(V_i^k, V_j^\ell) \quad \geqslant \quad \frac{|V_i\|V_j|}{n^2} \cdot d^2(V_i, V_j) + \frac{|V_i\|V_j|}{n^2} \cdot \epsilon^4.$$

$$(2.28)$$

Let π' be the common refinement of these partitions; say this partitions a set V_i as

$$\{V_{i,1}, \cdots, V_{i,k_i}\}.$$

Then $k_i \leqslant 2^k$ — so — $|\pi'| \leqslant k \cdot 2^k$.

By (2.28) and Lemma 2.49 we have for all irregular pairs $\{V_i, V_j\}$

$$\sum_{a=1}^{k_i} \sum_{b=1}^{k_j} \frac{|V_{i,a}||V_{j,b}|}{n^2} \cdot d^2(V_{i,a}, V_{j,b}) \geqslant$$

$$\frac{|V_i||V_j|}{n^2} \cdot d^2(V_i, V_j) + \frac{|V_i||V_j|}{n^2} \cdot \epsilon^4.$$

Since there are more than ϵk^2 irregular pairs — and since for $i \geqslant 1$: $|V_i| \geqslant \frac{(1-\epsilon)\cdot n}{k}$ — we find

$$\text{index}(\pi') \quad >$$

$$\text{index}(\pi) + \epsilon k^2 \cdot \frac{((1-\epsilon)n/k)^2}{n^2} \cdot \epsilon^4 \quad \geqslant \quad \text{index}(\pi) + \frac{1}{4} \cdot \epsilon^5$$

(since we may assume that $0 < \epsilon \leqslant 1/2$).

This proves the lemma.   □

WE NOW PROVE THE REGULARITY LEMMA (Lemma 2.47 on Page 55).

Proof. Start with an equitable partition $\pi_0 = \{V_0, \cdots, V_t\}$ with $|V_0| = t-1$. We may assume that n is large enough ie $t \leqslant \epsilon \cdot n/2$. [30]

If π_0 is not ϵ-regular there exists a refinement π' which satisfies

$$\text{index}(\pi') \quad \geqslant \quad \text{index}(\pi) + \frac{1}{4} \cdot \epsilon^5 \quad \text{and} \quad |\pi'| \quad \leqslant \quad |\pi| \cdot 2^{|\pi|}.$$

Let $A = |\pi'|$. To make π' into an equitable partition π_1 partition each class further into classes of size exactly $\lfloor \frac{1}{4} \cdot \epsilon^6 \cdot \frac{n}{A} \rfloor$ and at most one class of size less than that. All the small parts are moved into the exceptional set. This increases the size of the exceptional set by at most $\frac{1}{4} \cdot \epsilon^6 \cdot n$. [31] By Lemmas 2.49 and 2.51:

$$\text{index}(\pi_1) \quad \geqslant \quad \text{index}(\pi_0) + \frac{1}{4} \cdot \epsilon^5.$$

REPEATING THIS PROCESS the kth partition π_k satisfies

$$\text{index}(\pi_k) \quad \geqslant \quad \text{index}(\pi_0) + \frac{k}{4} \cdot \epsilon^5.$$

[30] This leaves us some space in the exceptional class — which we need — because the exceptional class grows during the refinements.

[31] That is so because the increase is less than

$$A \cdot \lfloor \epsilon^6 \cdot \frac{n}{4A} \rfloor \leqslant \frac{1}{4} \cdot \epsilon^6 \cdot n.$$

Notice that for all this to make sense we need $n > 4A/\epsilon^6$ otherwise the refinement does not exist. The iteration blows up this lower bound on n; but it remains constant.

However, any index is bounded from above by $\frac{1}{2}$. Therefore, the process ends in at most $2 \cdot \epsilon^{-5}$ iterations.

Notice that the increase of the exceptional set is smaller than

$$\frac{1}{4} \cdot \epsilon^6 \cdot n \cdot 2 \cdot \epsilon^{-5} \;=\; \frac{1}{2} \cdot \epsilon \cdot n,$$

— that is — since the original exceptional class satisfies

$$|V_0| \;\leqslant\; \frac{1}{2} \cdot \epsilon \cdot n,$$

we obtain an ϵ-regular partition. The number of classes is bounded by a function of ϵ and t.

This proves the regularity lemma.   □

Let a function $f : \mathbb{N} \to \mathbb{N}$ be defined by $f(1) = t$ and $f(k+1) = \left\lceil \frac{4}{\epsilon^6} \cdot f(k)2^{f(k)} \right\rceil$. The number of classes in the ϵ-regular partition π_k produced by the algorithm satisfies

$$|\pi_k| \;\leqslant\; f(\lceil 2\epsilon^{-5} \rceil).$$

Clearly, this is also the lower bound for n.

Remark 2.52. It is generally not possible to obtain ϵ-regular partitions that do not have any irregular pairs. The following bipartite graph is an example of a graph in which every ϵ-regular partition has irregular pairs. Let

$$A = \{a_1, \cdots, a_n\} \quad \text{and} \quad B = \{b_1, \cdots, b_n\}$$

and let $\{a_i, b_j\} \in E$ if $i \leqslant j$. (This bipartite graph is called a chain graph.)

2.7.1 Construction of regular partitions

Alon et al. showed (in 1994) that it is co-NP-complete to decide whether a given partition of a graph is ϵ-regular. — On the other hand — the lemma can be made constructive; an ϵ-regular partition can be found in $O(M(n))$ time, where $M(n) = n^{2.373}$ is the time needed to multiply two $n \times n$ matrices with entries in $\{0, 1\}$.

The problem to decide whether a given partition is ϵ-regular remains co-NP-complete even when $\epsilon = 1/2$ and $k = 2$. To prove that Alon et al. derive the following lemma (we omit the proof).

Lemma 2.53. *The following problem is NP-complete. Given a bipartite graph with color classes A and B satisfying $|A| = |B| = n$ and $|E| = \frac{1}{2} \cdot n^2 - 1$. Decide if the graph contains a complete bipartite subgraph $K_{\frac{n}{2}, \frac{n}{2}}$.*

To see that ϵ-regularity is co-NP-complete, Alon et al. make the following observation.

Lemma 2.54. *A bipartite graph with n vertices in each color class and with exactly $\frac{n^2}{2} - 1$ edges contains $K_{\frac{n}{2},\frac{n}{2}}$ if and only if it is <u>not</u> ϵ-regular with $\epsilon = 1/2$.*

Proof. Let the color classes be A and B. Assume $|A| = |B| = n$ and $d(A, B) = \frac{1}{2} - \frac{1}{n^2}$.

Assume the graph contains $K_{\frac{n}{2},\frac{n}{2}}$. Then there are sets X and Y of size $n/2$ and $d(X, Y) = 1$. This implies

$$| d(X, Y) - d(A, B) | \quad > \quad \frac{1}{2},$$

which shows that the graph is not ϵ-regular with $\epsilon = 1/2$.

Assume that the graph is not ϵ-regular with $\epsilon = 1/2$. Then there are sets X and Y of size at least $\frac{n}{2}$ that satisfy

$$| d(X, Y) - \frac{1}{2} + \frac{1}{n^2} | \quad > \quad \frac{1}{2}.$$

Notice that this is only possible when $d(X, Y) = 1$ — that is — $\{X, Y\}$ represents a complete bipartite subgraph (with at least $n/2$ vertices in each color class).

This proves the lemma.   □

In their paper Alon et al. prove the following theorem (which is a CONSTRUCTIVE EDITION of the Regularity Lemma (Lemma 2.47 on Page 55).

Theorem 2.55. *For $\epsilon > 0$ and $t \in \mathbb{N}$ there exist $N, T \in \mathbb{N}$ such that every graph with at least N vertices has an ϵ-regular partition with $k + 1$ classes where k satisfies $t \leqslant k \leqslant T$.*

Such a partition can be found in $O(M(n))$ time.

To prove the theorem we need to do some preliminary work.

Definition 2.56. Let A and B be color classes of a bipartite graph satisfying $|A| = |B| = n$. Let d be the average degree. For two vertices $p, q \in B$ define their <u>neighborhood deviation</u> as

$$\sigma(p, q) = |N(p) \cap N(q)| - \frac{d^2}{n}.$$

For $Y \subseteq B$ ($Y \neq \varnothing$) define its deviation as [32]

$$\sigma(Y) = \frac{1}{|Y|^2} \cdot \sum_{\substack{p,q \in Y \\ p \neq q}} \sigma(p, q).$$

A constructive version of the regularity lemma depends on the construction of suitable subsets V_i' and V_j' satisfying (2.26) (of irregular pairs $\{V_i, V_j\}$). These 'witnesses' to the irregularity are hard to find.

The following lemma shows that it is possible to 'approximate' witnesses.

Lemma 2.57. *Let a bipartite graph have color classes A and B which satisfy* $|A| = |B| = n$. *Let* d *be the average degree. Let* $0 < \epsilon < 1/16$. *Assume there exists a set* $Y \subseteq B$ *such that*

$$|Y| \geqslant \epsilon \cdot n \quad and \quad \sigma(Y) \geqslant \frac{\epsilon^3}{2} \cdot n \qquad (2.29)$$

Then one of the following items holds true

I. $d < \epsilon^3 \cdot n$;

II. $|\{y \mid y \in B \quad and \quad |\deg(y) - d| \geqslant \epsilon^4 \cdot n\}| > \frac{\epsilon^4}{8} \cdot n$;

III. *There exist subsets* $A' \subseteq A$ *and* $B' \subseteq B$ *that satisfy*

$$|A'| \geqslant \frac{\epsilon^4}{4} \cdot |A| \quad and \quad |B'| \geqslant \frac{\epsilon^4}{4} \cdot |B| \quad and$$

$$|d(A', B') - d(A, B)| > \epsilon^4.$$

[32] When the graph is complete bipartite or empty then

$$\sigma(p, q) = 0$$

for all pairs $p, q \in B$. In that case $\sigma(Y) = 0$ for all nonempty sets $Y \subseteq B$.

As another example, define a bipartite graph H on the 7 points and 7 lines of the Fano plane. A point and line are adjacent of they are incident in the plane. We have an average degree $d = 3$ since any point lies on 3 lines. Since any pair of points lie on one line

$$\sigma(p, q) = 1 - 9/7 = -\frac{2}{7}.$$

When B is the set of points of the Fano plane, we find

$$\sigma(B) = \frac{1}{49} \cdot 7 \cdot 6 \cdot \left(-\frac{2}{7}\right) = -\frac{12}{49}.$$

There exists an $O(M(n))$-algorithm that produces sets A' and B' as specified in III *when* I *and* II *do not hold.*

Proof. Assume that I and II do not happen. We prove III.

Define

$$Y' \;\; - \;\; \{y \in Y \;\; | \;\; |dog(y) \;\; d| \;\; < \;\; \varepsilon^4 \cdot n\}.$$

Then $Y' \neq \varnothing$ since II does not occur.

Choose $y_0 \in Y$ that maximizes $\sum_{y \in Y} \sigma(y, y_0)$. By taking an average we estimate $\sum_{y \in Y} \sigma(y, y_0)$.

$$\sum_{\substack{y' \in Y' \\ y \neq y'}} \sum_{y \in Y} \sigma(y, y') = \sigma(Y) \cdot |Y|^2 \;\; - \sum_{\substack{y' \in Y \setminus Y' \\ y \neq y'}} \sum_{y \in Y} \sigma(y, y')$$

$$\geqslant \;\; \frac{\varepsilon^3}{2} \cdot n \cdot |Y|^2 \;\; - \;\; \frac{\varepsilon^4}{8} \cdot n \cdot |Y| \cdot n.$$

(Here we use Y's property that $\sigma(Y) \geqslant \varepsilon^3 \cdot n / 2$ and the assumption that II does not occur which implies $|Y \setminus Y'| \leqslant \frac{\varepsilon^4}{8} \cdot n$.)
Since $|Y'| \leqslant |Y|$ and $|Y| \geqslant \varepsilon \cdot n$, we have

$$\sum_{y \in Y} \sigma(y, y_0) \;\; \geqslant \;\; \frac{\varepsilon^3}{2} \cdot n \cdot |Y| - \frac{\varepsilon^4}{8} \cdot n^2 \;\; \geqslant \;\; \frac{3}{8} \cdot \varepsilon^3 \cdot n \cdot |Y|. \quad (2.30)$$

We CLAIM

$$| \{y \;\; | \;\; y \in Y \;\; \text{and} \;\; \sigma(y, y_0) \;\; > \;\; 2\varepsilon^4 \cdot n\} | \;\; \geqslant \;\; \frac{\varepsilon^4}{4} \cdot n. \quad (2.31)$$

OTHERWISE

$$\sum_{y \in Y} \sigma(y, y_0) \;\; < \;\; \frac{\varepsilon^4}{4} \cdot n^2 + |Y| \cdot 2\varepsilon^4 \cdot n$$

$$\leqslant \;\; \frac{\varepsilon^3}{4} \cdot n \cdot |Y| + 2\varepsilon^4 \cdot n \cdot |Y| \;\; \leqslant \;\; \frac{3}{8} \cdot \varepsilon^3 \cdot n \cdot |Y|$$

and this contradicts (2.30).

It follows from (2.31) that there exists a set $B' \subseteq Y \setminus \{y_0\}$ which satisfies

$$|B'| \geqslant \frac{\varepsilon^4}{4} \cdot n \quad \text{and} \quad \forall_{b \in B'} \;\; |N(b) \cap N(y_0)| \;\; > \;\; \frac{d^2}{n} + 2\varepsilon^4 \cdot n.$$

Let $A' = N(y_0)$. Since $d \geqslant \epsilon^3 \cdot n$ and $16\epsilon < 1$,

$$|A'| \quad \geqslant \quad d - \epsilon^4 \cdot n \quad \geqslant \quad \epsilon^3 \cdot n - \epsilon^4 \cdot n \quad \geqslant \quad 15\epsilon^4 \cdot n \quad \geqslant \quad \frac{\epsilon^4}{4} \cdot n.$$

We CLAIM
$$|d(A', B') - d(A, B)| \quad > \quad \epsilon^4.$$

TO SEE THAT notice

$$e(A', B') = \sum_{b \in B'} |N(y_0) \cap N(b)| \quad > \quad \left(\frac{d^2}{n} + 2\epsilon^4 \cdot n \right) \cdot |B'|.$$

It follows that

$$d(A', B') - d(A, B) \quad > \quad \left(\frac{d^2}{n} + 2\epsilon^4 \cdot n \right) \cdot \frac{1}{|A'|} - \frac{d}{n}$$

$$> \quad \frac{d^2}{n(d + \epsilon^4 \cdot n)} + 2\epsilon^4 - \frac{d}{n} \quad =$$

$$2\epsilon^4 \quad - \quad \frac{d\epsilon^4}{d + \epsilon^4 \cdot n} \quad \geqslant \quad \epsilon^4.$$

(In the second line we use the fact that $|A'| \leqslant n$ since $A' \subseteq A$ and the fact that $|A'| = |N(y_0)| < d + \epsilon^4 \cdot n$ since $y_0 \in Y'$.)

AN ALGORITHM to compute the sets A' and B' proceeds by computing for all $y_0 \in B$ with $|\deg(y_0) - d| < \epsilon^4 \cdot n$ the set

$$B(y_0) \quad = \quad \{y \mid y \in B \text{ and } \sigma(y, y_0) > 2\epsilon^4 \cdot n\}.$$

There exists a vertex y_0 with $|B(y_0)| \geqslant \frac{\epsilon^4}{4} \cdot n$ (see (2.31)). The required sets are $B' = B(y_0)$ and $A' = N(y_0)$.

All quantities $\sigma(y, y')$ can be obtained by squaring the adjacency matrix.

This proves the lemma.   □

Exercise 2.31

Let d be the average degree of a bipartite graph with color classes A and B of size n. Assume $d < \epsilon^3 \cdot n$ (Case I of Lemma 2.57). Prove that the graph is ϵ-regular.

HINT: See Exercise 2.27.

Exercise 2.32

Let the color classes A and B of a bipartite graph satisfy $|A| = n$ and $|B| = n$. Show that the graph is ϵ-regular if for all $X \subseteq A$ and $Y \subseteq B$

$$|X| = \lceil \epsilon \cdot |A| \rceil \quad \text{and} \quad |Y| = \lceil \epsilon \cdot |B| \rceil \quad \Rightarrow$$
$$|d(X, Y) - d(A, B)| \quad \leqslant \quad \epsilon.$$

HINT: Compare Definition 2.44 (which is on Page 54).

Lemma 2.58. *Let A and B be color classes of a bipartite graph and assume that $|A| = |B| = n$. Assume*

$$2n^{-\frac{1}{4}} \quad < \quad \epsilon \quad < \quad \frac{1}{16}.$$

Assume that case II, *in Lemma 2.57, does not occur, ie*

$$|\{y \in B \mid |\deg(y) - d| \quad \geqslant \quad \epsilon^4 \cdot n\}| \quad \leqslant \quad \frac{\epsilon^4}{8} \cdot n.$$

If the graph is <u>*not*</u> *ϵ-regular then there exists a set $Y \subset B$ which satisfies* (2.29) *ie*

$$|Y| \quad \geqslant \quad \epsilon \cdot n \quad \text{and} \quad \sigma(Y) \quad \geqslant \quad \frac{\epsilon^3}{2} \cdot n. \qquad (2.32)$$

Proof. Assume that no set $Y \subseteq B$, $|Y| \geqslant \epsilon \cdot n$ satisfies (2.32). We show that the graph is ϵ-regular.

Let $X \subseteq A$ and $Y \subseteq B$ and let

$$|X| = \lceil \epsilon \cdot |A| \rceil \quad \text{and} \quad |Y| = \lceil \epsilon \cdot |B| \rceil.$$

We show that
$$|d(X, Y) - d(A, B)| \quad \leqslant \quad \epsilon.$$

We CLAIM

$$\sum_{x \in X} \left(|N(x) \cap Y| - \frac{d}{n} \cdot |Y| \right)^2 \quad \leqslant \quad e(A, Y) + |Y|^2 \cdot \sigma(Y) + \frac{2}{5} \epsilon^5 \cdot n^3.$$

TO SEE THAT write $(m_{i,j})$ for the bipartite adjacency matrix. Then

$$\sum_{x \in X} \left(|N(x) \cap Y| - \frac{d}{n} \cdot |Y| \right)^2 \quad \leqslant$$

$$\sum_{x \in A} \left(|N(x) \cap Y| - \frac{d}{n} \cdot |Y| \right)^2 \quad =$$

$$\sum_{x \in A} \left(\sum_{y \in Y} m_{x,y} - \frac{d}{n} \cdot |Y| \right)^2 \quad = \quad \sum_{x \in A} \left(\sum_{y \in Y} m_{x,y}^2 + \frac{d^2}{n^2} \cdot |Y|^2 + \right.$$

$$\left. \sum_{\substack{y,y' \in Y \\ y \neq y'}} m_{x,y} \cdot m_{x,y'} - \frac{2d}{n} \cdot |Y| \cdot \sum_{y \in Y} m_{x,y} \right) \quad =$$

$$e(A, Y) + \frac{d^2}{n} \cdot |Y|^2 + \sum_{\substack{y,y' \in Y \\ y \neq y'}} |N(y) \cap N(y')| - \frac{2d}{n} \cdot |Y| \cdot e(A, Y) \quad =$$

$$e(A, Y) + \frac{d^2}{n} \cdot |Y|^2 + \sum_{\substack{y,y' \in Y \\ y \neq y'}} \left(\sigma(y, y') + \frac{d^2}{n} \right) - \frac{2d}{n} \cdot |Y| \cdot e(A, Y) \quad \leqslant$$

$$e(A, Y) + \sigma(Y) \cdot |Y|^2 + \frac{2d^2}{n} \cdot |Y|^2 - \frac{2d}{n} \cdot |Y| \cdot e(A, Y).$$

To prove the claim, we must show

$$\frac{d^2}{n} \cdot |Y|^2 - \frac{d}{n} \cdot |Y| \cdot e(A, Y) \quad \leqslant \quad \frac{\epsilon^5}{5} \cdot n^3,$$

that is,

we need to show that:
$$\boxed{d(A, Y) \quad \geqslant \quad \frac{d}{n} - \frac{\epsilon^5 \cdot n^3}{5d} \cdot \frac{1}{|Y|^2}.}$$

To see that notice

$$d(A, Y) = \frac{e(A, Y)}{n \cdot |Y|} \geqslant \frac{(d - \epsilon^4 \cdot n)(|Y| - \frac{\epsilon^4}{8} \cdot n)}{n \cdot |Y|} \quad =$$

$$\frac{d}{n} - \epsilon^4 - \frac{\epsilon^4 \cdot d}{8 \cdot |Y|} + \frac{\epsilon^8 \cdot n}{8 \cdot |Y|} \quad \geqslant \quad \frac{d}{n} - \epsilon^4 - \frac{\epsilon^3}{8}.$$

(We used $|Y| \geqslant \epsilon \cdot n \geqslant \epsilon \cdot d$.)

We now use the assumption that $\epsilon^4 \cdot n > 1$ and that $|Y| \leqslant \epsilon \cdot n + 1$:

$$\frac{\epsilon^5 \cdot n^3}{5d \cdot |Y|^2} \quad \geqslant \quad \frac{\epsilon^5 \cdot n^2}{5(\epsilon \cdot n + 1)^2} \quad \geqslant \quad \frac{\epsilon^5}{5(\epsilon + \epsilon^4)^2} \quad \geqslant \quad \frac{\epsilon^3}{8} + \epsilon^4.$$

This proves the claim.

By the Cauchy-Schwartz inequality we have

$$\sum_{x \in X} \left(|N(x) \cap Y| - \frac{d}{n} \cdot |Y| \right)^2 \geqslant$$

$$\frac{1}{|X|} \left(\left(\sum_{x \in X} |N(x) \cap Y| \right) - \frac{d}{n} \cdot |X| \cdot |Y| \right)^2 ,$$

— So — by the previous claim

$$\left(\left(\sum_{x \in X} |N(x) \cap Y| \right) - \frac{d}{n} \cdot |X| \cdot |Y| \right)^2 \leqslant$$

$$|X| \cdot \left(e(A, Y) + |Y|^2 \cdot \sigma(Y) + \frac{2\epsilon^5}{5} \cdot n^3 \right).$$

When we divide by $|X|^2 \cdot |Y|^2$ we obtain

$$|d(X, Y) - d(A, B)| \leqslant$$

$$\frac{1}{|X| \cdot |Y|} \cdot \left(e(A, Y) + |Y|^2 \cdot \sigma(Y) + \frac{2\epsilon^5}{5} \cdot n^3 \right).$$

We now use that

(i) $e(A, Y) \leqslant (d + \epsilon^4 \cdot n) \cdot |Y| + \frac{\epsilon^4}{8} \cdot n^2$

(ii) $\sigma(Y) \leqslant \frac{\epsilon^3}{2} \cdot n$

(iii) $\epsilon > 2n^{-1/4}$,

and we find

$$|d(X, Y) - d(A, B)|^2 \leqslant$$

$$\frac{1}{|X| \cdot |Y|^2} \cdot \left((d + \epsilon^4 \cdot n) \cdot |Y| + \frac{\epsilon^4 \cdot n^2}{8} + \frac{\epsilon^3 \cdot n}{2} \cdot |Y|^2 + \frac{2\epsilon^5}{5} \cdot n^3 \right) \leqslant$$

$$\frac{n + \epsilon^4 \cdot n}{\epsilon^2 \cdot n^2} + \frac{\epsilon^4 \cdot n^2}{8\epsilon^3 \cdot n^3} + \frac{\epsilon^3 \cdot n}{2\epsilon \cdot n} + \frac{2\epsilon^5 \cdot n^3}{5\epsilon^3 \cdot n^3} \leqslant$$

$$\frac{\epsilon^2 \cdot (1 + \epsilon^4)}{16} + \frac{9\epsilon^2}{10} + \frac{\epsilon^5}{128} \leqslant \epsilon^2.$$

This proves that the graph is ϵ-regular — that is — it proves the lemma. □

Corollary 2.59. *Let a bipartite graph* H *have color classes* A *and* B *that satisfy*

$$|A| \quad = \quad |B| \quad = \quad n.$$

Assume

$$2 \cdot n^{-1/4} \quad < \quad \epsilon \quad < \quad \frac{1}{16}.$$

There exists an algorithm that runs in $O(M(n))$ *time and that verifies that* H *is* ϵ-*regular or finds sets* $A' \subseteq A$ *and* $B' \subseteq B$ *that satisfy*

$$|A'| \geqslant \frac{\epsilon^4}{16} \cdot n, \quad |B'| \geqslant \frac{\epsilon^4}{16} \cdot n, \quad |d(A', B') - d(A, B)| \geqslant \epsilon^4. \ (2.33)$$

Proof. If $d \leqslant \epsilon^3 \cdot n$, then we are done; we can check that in $O(n^2)$ time and report that H is ϵ-regular.

NEXT if

$$|\{y \mid y \in B \quad |\deg(y) - d| \geqslant \epsilon^4 \cdot n\}| \quad \geqslant \quad \frac{\epsilon^4}{8} \cdot n,$$

then the degrees of at least half of them deviate from d in the same direction. — That is — we can find a set $B' \subseteq B$ that satisfies $|B'| \geqslant \frac{\epsilon^4}{16} \cdot n$ and that satisfies

$$|d(A, B') - d(A, B)| \quad \geqslant \quad \epsilon^4.$$

In that case we are done.

FINALLY, we need to show that we can find A' and B' as required when H is not ϵ-regular — that is (by Lemma 2.58) —when there exists a set Y that satisfies Equation 2.29 ie

$$|Y| \quad \geqslant \quad \epsilon \cdot n \quad \text{and} \quad \sigma(Y) \quad \geqslant \quad \frac{\epsilon^3}{2} \cdot n. \quad (2.34)$$

By Lemma 2.57 there exists an $O(M(n))$ algorithm to compute these sets A' and B' when Y exists.

This proves the claim. $\square$

Alon et al proceed to formulate the key-Lemma 2.51 (on Page 58) in a constructive form.

Lemma 2.60. *Let* $k \in \mathbb{N}$ *and let* $0 < \gamma \leqslant 1/2$. *Let* $\Pi_0 = \{C_0, \cdots, C_k\}$ *be an equitable partition and assume*

$$|C_1| \ > \ 4^{2k} \quad and \quad |C_0| \ \leqslant \ \gamma \cdot n. \qquad (2.35)$$

Assume that proofs are given as part of the input of more than $\gamma \cdot k^2$ *pairs of classes showing that they are not* γ*-regular.* [33] *There is an algorithm that runs in* $O(n)$ *time that produces a refinement* Π_1 *of* Π_0 *satisfying*

$$|\Pi_1| \ = \ k \cdot 4^k.$$

Furthermore, the exceptional class increases by at most $n/2^k$ *and*

$$\mathsf{index}(\Pi_1) \ > \ \mathsf{index}(\Pi_0) + \frac{1}{4} \cdot \gamma^5.$$

[33] By 'proofs' we mean 'witnesses' ie subsets of the color classes that violate γ-regularity.

Proof. By Lemma 2.51 there is a linear-time algorithm that computes a refinement Π' of Π_0 (leaving the exceptional class unaltered but which may have other classes of unequal size) satisfying

$$|\Pi'| \ \leqslant \ k \cdot 2^k \quad and \quad \mathsf{index}(\Pi') \ > \ \mathsf{index}(\Pi_0) + \frac{1}{4} \cdot \gamma^5.$$

Since $|C_1| > 4^{2k}$ we can refine Π' into an equitable partition Π_1 which satisfies [34] (by Lemma 2.49)

$$|\Pi_1| \ = \ k \cdot 4^k \quad and \quad \mathsf{index}(\Pi_1) \ > \ \mathsf{index}(\Pi_0) + \frac{1}{4} \cdot \gamma^5.$$

This increases the size of the exceptional class of Π_0 (and of Π') by at most $n/2^k$. $\qquad\square$

[34] To obtain Π_1 partition the parts of Π' further into parts of size

$$\lfloor (n-|C_0|)/k \cdot 4^k \rfloor = \lfloor |C_1|/4^k \rfloor.$$

This increases the exceptional class by at most

$$|\Pi'| \cdot \left\lfloor \frac{n - |C_0|}{k \cdot 4^k} \right\rfloor \ \leqslant \ \frac{n}{2^k}.$$

BELOW FOLLOWS THE PROOF OF THEOREM 2.55:

Theorem. *For* $\epsilon > 0$ *and* $t \in \mathbb{N}$ *there exist* $N, T \in \mathbb{N}$ *such that every graph with at least* N *vertices has an* ϵ-regular partition with $k + 1$ classes where k satisfies $t \leqslant k \leqslant T$.

Such a partition can be found in $O(M(n))$ *time.*

Proof. Let $\epsilon > 0$ and $t \in \mathbb{N}$.

The following algorithm computes a sequence of equitable partitions — say (Π_i). Write $|\Pi_i| = k_i$ and let n_i be the size of a class of Π_i that is not exceptional. Let $\gamma = \epsilon^4/16$.

1. Start with an equitable partition $\Pi_1 = \{C_0, \cdots, C_{k_1}\}$ with

$$|C_0| \quad < \quad k_1 \quad \text{and} \quad |C_i| \quad = \quad n_1 \quad = \quad \lfloor n/k_1 \rfloor \quad \text{for } i > 0.$$

   (Below we determine a suitable value for k_1.) Set $i := 1$.

2. To proceed from a partition Π_i use Corollary 2.59 to determine for every pair of classes in P_i whether they are ϵ-regular or else to find pairs of subsets (of size at least $\frac{\epsilon^4}{16} \cdot n_i$) as in Equation 2.33. [35]

3. If at most $\epsilon \cdot k_i$ pairs are not ϵ-regular then report Π_i as an ϵ-regular partition. STOP.

4. Otherwise call on Lemma 2.60 to compute a refinement Π_{i+1} of Π_i satisfying

$$k_{i+1} \quad = \quad k_i \cdot 4^{k_i} \quad \text{and} \quad \text{index}(\Pi_{i+1}) \quad \geqslant \quad \text{index}(\Pi_i) + \frac{1}{4} \cdot \gamma^5.$$

   Write C_0^i for the exceptional class of Π_i. [36] Then the exceptional class of Π_{i+1} satisfies

$$|C_0^{i+1}| - |C_0^i| \quad \leqslant \quad \frac{n}{2^{k_i}}.$$

   Notice that to apply Lemma 2.60 we need to ensure Equation 2.35 — ie —

$$n_i \quad > \quad 4^{2k_i} \quad \text{and} \quad |C_0^i| \quad \leqslant \quad \gamma \cdot n.$$

5. Set $i := i + 1$ and go to Step 2.

NOTICE THAT

$$\text{index}(\Pi_i) \quad \geqslant \quad \text{index}(\Pi_1) + (i - 1) \cdot \frac{\gamma^5}{4}.$$

So — since the index cannot exceed $1/2$ — the algorithm ends in at most $1 + \lfloor 2 \cdot \gamma^{-5} \rfloor$ iterations.

---

[35] To apply the corollary we need to ensure that n is large enough to satisfy the condition $\epsilon^4 \cdot n_i > 16$ ie $\gamma \cdot n_i > 1$.

[36] To compute Π_{i+1} use Lemma 2.51 to compute Π' (on input Π_i) with

$$|\Pi'| \leqslant k_i \cdot 2_i^k$$

Refine Π' into parts of size

$$n_{i+1} = \left\lfloor \frac{n - |C_0^i|}{k_i \cdot 4^{k_i}} \right\rfloor.$$

This increases the exceptional class

$$|C_0^{i+1}| - |C_0^i| \leqslant$$
$$k_i \cdot 2^{k_i} \cdot \left\lfloor \frac{n - |C_0^i|}{k_i \cdot 4^{k_i}} \right\rfloor \leqslant \frac{n}{2^{k_i}}$$

To define a lowerbound for the number of vertices in the graph introduce a function $f : \mathbb{N} \to \mathbb{N}$ as follows

$$f(i) \quad = \quad \begin{cases} k_1 & \text{if } i = 1 \\ f(i-1) \cdot 4^{f(i-1)} & \text{otherwise.} \end{cases}$$

Let $T = f\left(1 + \lfloor 2 \cdot \gamma^{-5} \rfloor\right)$, let $N = \max\{ 2 \cdot T \cdot 4^{2 \cdot T}, \frac{2 \cdot T}{\gamma} \}$ and assume that $n \geqslant N$.

We show that the exceptional class does not exceed $\gamma \cdot n$. We may assume that

$$\gamma \cdot 2^{k_1} \quad > \quad 4. \tag{2.36}$$

CLAIM: $|C_0^i| \leqslant \gamma \cdot n \cdot (1 - 1/2^i)$. For $i = 1$ we have

$$n \quad \geqslant \quad N \quad \geqslant \quad \frac{2 \cdot T}{\gamma} \quad \geqslant \quad \frac{2 \cdot k_1}{\gamma} \quad \Rightarrow$$

$$|C_0^1| \quad < \quad k_1 \quad \leqslant \quad \frac{\gamma \cdot n}{2}.$$

This show that the claim is true for $i = 1$.

We have

$$|C_0^{i+1}| - |C_0^i| \quad \leqslant \quad \frac{n}{2^{k_i}} \quad \text{and we show:} \quad \frac{n}{2^{k_i}} \quad < \quad \gamma \cdot \frac{n}{2^{i+1}}.$$

This is true for $i = 1$ by the assumption (2.36) and

$$k_{i+1} = k_i \cdot 4^{k_i} \quad \Rightarrow \quad \gamma \cdot 2^{k_{i+1}} \quad > \quad 2 \cdot \gamma \cdot 2^{k_i} \quad \geqslant \quad \gamma \cdot 2^{i+1}.$$

This proves the claim.

Corollary 2.59 requires that

$$\gamma \cdot n_i \quad > \quad 1.$$

To see that this holds true observe

$$n_i \quad \geqslant \quad \frac{(1 - \gamma) \cdot n}{k_i} \quad \geqslant \quad \frac{(1 - \gamma) \cdot N}{T} \quad \geqslant \quad 2 \cdot \frac{1 - \gamma}{\gamma}.$$

Since $\gamma < 1/2$ this implies

$$\gamma \cdot n_i \quad \geqslant \quad 2 \cdot (1 - \gamma) \quad > \quad 1.$$

Lemma 2.60 requires that

$$n_i \; > \; 4^{2k_i}.$$

We have

$$n_i \geqslant {}^{(1-\gamma)\cdot N}/k_i \quad \text{and} \quad k_i \leqslant T \quad \text{and} \quad N \geqslant 2\cdot T\cdot 4^{2\cdot T}.$$

It easily follows that

$$\frac{n_i}{4^{2\cdot k_i}} \;\; \geqslant \;\; 2\cdot(1-\gamma) \;\; > \;\; 1.$$

This proves the theorem.   □

Remark 2.61. In 2014 J. Fox and L. M. Lovász proved tight lowerbounds for the number of parts in an ϵ-regular partition.

2.8   *Edge - thickness and stickiness*

For a set S let

$$\Lambda(S) \;\; = \;\; \{\lambda : S \to \mathbb{R}^{\geqslant 0} \mid \sum_{x\in S} \lambda(x) = 1\}.$$

The elements of $\Lambda(S)$ are called thickness functions. For a thickness function λ and $A \subseteq S$ write $\lambda(A) = \sum_{x\in A} \lambda(x)$.

Definition 2.62. Let G be a graph. The <u>thickness</u> of G is defined as

$$\phi(G) \;\; = \;\; \inf_{\lambda\in\Lambda(V)} \; \sup_{e\in E} \; \lambda(e).$$

T. Kloks, C. Lee and J. Liu, *Stickiness, edge-thickness, and clique-thickness in graphs*, Journal of Information Science and Engineering **20** (2004), pp. 207–217.

The inf and sup may be replaced by min and max.

Exercise 2.33

Assume that G is a graph which is not empty. Show that

$$\frac{\delta}{m} \;\; \leqslant \;\; \phi(G) \;\; \leqslant \;\; \frac{1}{\alpha(G)},$$

where m is the number of edges and δ is the minimal degree of a vertex in G and $\alpha(G)$ is the size of a largest independent set in G.

Exercise 2.34

(i) CLEARLY when G has an isolated vertex then its thickness is 0.

(ii) Assume that G has no isolated vertices. Let λ be a thickness function that realizes the thickness of G. Denote $\phi = \phi(G)$. Show that every vertex is incident with an edge of weight ϕ.

(iii) Assume that G has no isolated vertices. Show that there exists a thickness function λ that realizes the thickness of G and that has a range

$$\{\, 0, \phi, \frac{\phi}{2} \,\}.$$

Definition 2.63. The stickiness of a graph G is

$$s(G) \quad = \quad \max_{X \subseteq V} \quad |X| - |\bar{N}(X)|$$

$$\text{where} \quad \bar{N}(X) \quad = \quad \cup_{x \in X} N(x).$$

A graph is Hallian if $s(G) \leqslant 0$. A graph is Hallian if and only if there is a partition of V into a matching and a set of odd cycles.

Exercise 2.35

When $s(G) > 0$ then it is realized by an independent set.

Exercise 2.36

If G has an isolated vertex then $\phi(G) = 0$ otherwise

$$\phi(G) \quad = \quad \frac{2}{n + s(G)}.$$

Computing stickiness

Let G be a graph. Construct a bipartite graph G' as follows. Create two copies of a vertex $x \in V$ — say x^1 and x^2. For $A \subseteq V$ write $A^i = \{x^i \mid x \in A\}$. The set of vertices of G' is $V^1 \cup V^2$. The edges of G' are

$$E(G') \quad = \quad \{\{x^1, y^2\} \mid \{x, y\} \in E(G)\}.$$

Construct a flow - network on G' by giving each edge of G' a capacity ∞. Add one source - vertex s and make it adjacent to all vertices of V^1. All edges that are incident with s have capacity 1. Add one sink - vertex t and make it adjacent to all vertices of V^2. All edges incident with t have capacity 1.

J. Orlin, *Max flows in* $O(nm)$ *time, or better*, Proceedings STOC'13, pp. 765–774.

A cut in the network is a set of the form $A^1 \cup B^2 \cup \{s\}$ where $A \subseteq V$ and $\bar{N}(A) \subseteq B$. The capacity of such a cut is

$$|V \setminus A| \quad + \quad |B| \quad = \quad n \quad + \quad |B| \quad - \quad |A|.$$

A cut of minimum capacity satisfies $B = \bar{N}(A)$ — and so — it minimizes $|\bar{N}(A)| - |A|$.

A minimum cut can be computed $O(n \cdot m)$ time.

Remark 2.64 (Motzkin - Straus Theorem (1965)). Let G be a graph with vertex set $[n]$. Give each vertex i a values $x_i \geqslant 0$ such that $\sum_i x_i = 1$. The largest value over these assignments of

$$\sum_{(i,j) \in E} x_i \cdot x_j$$

is equal to $\frac{1}{2} \cdot (1 - 1/\omega)$.

2.9   Clique Separators

Definition 2.65. Let G be a graph. A set $S \subseteq V$ is a clique separator if $S = \varnothing$ or a clique and $G - S$ is disconnected.

Whitesides showed that one can find a clique cutset in $O(n^3)$ time. In this section we present a variation of this algorithm. We show that there exists an $O(n^4)$ algorithm that lists all minimal clique separators.

Definition 2.66. Let G be a graph. A set $S \subseteq V$ is a minimal clique separator if S is a minimal separator — and — either $S = \varnothing$ or a clique.

Example 2.67. Notice that the number of clique separators that a graph may have is exponential. — For example — consider the graph built up from a clique K_t and a path P_3, say $[x, y, z]$. Add additional edges from the midpoint y to all vertices of the clique K_t. Then, for any subset $W \subseteq V(K_t)$

$$W \cup [y] \quad \text{is a clique separator.}$$

This shows that the graph has at least 2^t clique separators, and $t + 3$ vertices.

On the other hand the graph has only one minimal separator — namely — $\{y\}$.

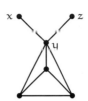

Figure 2.6: This graph has an exponential number of clique separators, as the size of the clique goes to ∞.

For a graph G, write $\sigma(G)$ for the number of minimal clique separators in G.

Lemma 2.68. *For any graph* $\sigma < |V|$.

Proof. We claim that, when G is connected, $\sigma(G) < n$. Notice that this implies the lemma. — In order to prove this — we may assume that G is connected. We may also assume that G is not a clique — as otherwise $\sigma(G) = 0 < n$.

Let $x \in V$ be a vertex for which the largest component in $G - N[x]$ is as large as possible. Let C be this largest component. Let

$$S = N(C) \quad \text{and} \quad X = V \setminus N[C]. \tag{2.37}$$

Then $x \in X$ — so $X \neq \varnothing$. Since G is connected $S \neq \varnothing$, and since G is not a clique $C \neq \varnothing$ — so $\{X, S, C\}$ is a partition of the set V.

— By our choice of x — every vertex of X is adjacent to every vertex of S. This implies that every minimal clique separator in $G[X \cup S]$ either contains X or else contains S. — So —

Here, we let a set stand in for the graph induced by that set.

$$\sigma(X \cup S) = \begin{cases} \sigma(X) & \text{if } S \text{ is a clique and } X \text{ is not} \\ \sigma(S) & \text{if } X \text{ is a clique and } S \text{ is not} \\ 0 & \text{if neither or both are cliques.} \end{cases} \tag{2.38}$$

Any vertex of S has a neighbor in C, and since C is connected, no two vertices in S are separated by a clique separator. This reduces the number (2.38) of minimal clique separators in $X \cup S$ that are clique separators in G, to at most $\sigma(X)$.

A minimal clique separator which is contained in $C \cup S$ may no longer be a minimal separator in $G - X$. — To repair that — in the graph $C \cup S$, add all edges between pairs in S.[37] Denote this graph as $C \cup \bar{S}$.

Some minimal clique separators in $C \cup \bar{S}$ may not be cliques in G [38] — but — as an upperbound for the number of minimal clique separators — that are contained in $C \cup S$ — we find $\sigma(C \cup \bar{S})$.

Using induction on the number of vertices in the graph we find:

$$\begin{aligned} \sigma(G) &\leqslant \sigma(X) + \sigma(C \cup \bar{S}) \\ &< |X| + |C \cup S| \quad (\text{since } C \cup S \text{ is connected}) \\ &= n. \end{aligned}$$

This proves the lemma.   □

[37] A clique separator T, contained in $C \cup S$, has all vertices of S that are not in T in one component of $G - T$. Removal of the set X may disconnect this component and possibly T is not a minimal separator in $G - X$.

[38] In our algorithm, we will sift the minimal clique separators of $C \cup \bar{S}$, and select those that are cliques in G.

Exercise 2.37

Prove the first inequality. Hint: Show that every minimal clique separator is one in $G[X \cup S]$ or one in $G[C \cup S]$.

2.9.1   Feasible Partitions

To turn the proof of Lemma 2.68 into an algorithm we need to find the partition $\{X, S, C\}$ as mentioned there. — That is — we need a feasible partition of $V(G)$ — as defined below.

Definition 2.69. A partition $\{X, S, C\}$ of V is <u>feasible</u> if

(a) $G[C]$ is connected and

(b) $S = N(C)$ and S separates X and C and

(c) every vertex of X is adjacent to every vertex of S.

In this section we show how to find a feasible partition. The idea is to start with $C = \{R\}$, for some arbitrary vertex $R \in V$ which is not universal.[39]

[39] A vertex is <u>universal</u> if it it adjacent to all others.

As an invariant for the algorithm we let $\{X, S, C\}$ be a partitition of V with the property

C is connected and $S = N(C)$ and S separates X and C.   (2.39)

Notice that the partition $\{X, S, C\}$ satisfies (2.39) at the initialization — when $C = \{R\}$ — since R is nonuniversal.[40]

[40] If every vertex is universal, then G is a clique, and then, there is no minimal separator.

TO MAKE PROGRESS — towards a feasible partition — let C grow as follows:

$$\text{Choose } y \in S \text{ with } X \not\subset N(y).   \qquad (2.40)$$

The vertex y is added to C.

Notice that, when such a vertex $y \in S$ does not exist, we may conclude the <u>postcondition</u> — that is — $\{X, S, C\}$ is a feasible partition.

The weakest precondition is a property that ensures progress when valid, and that ensures the postcondition when not valid. In our case, the weakest precondition is the existence of a vertex y as in (2.40).

This proves the correctness of Algorithm 5.

Exercise 2.38

Let $T(n)$ denote the worst–case running-time bound of algorithm 5. Prove that

$$T(n) \leqslant T(n-1) + O(n^2) \quad \Rightarrow \quad T(n) = O(n^3). \qquad (2.41)$$

```
1: procedure  FP ( G )
2:
3:     if G is a clique then
4:         there is no partition
5:     else
6:         R ←∈ { x | x ∈ V   and   d(x) < n − 1 }
7:         C ← {R}
8:         S ← N(R)
9:         X ← V \ N[C]
10:
11:        while ∃_{y∈S} X ⊄ N(y)  do
12:            C ← C ∪ {y}
13:            S ← ( S \ {y} ) ∪ ( N(y) ∩ X )
14:            X ← X \ N(y)
15:        end while
16:
17:        report { X, S, C }
18:     end if
19:
20: end procedure
```

2.9.2 Intermezzo

Another way to describe the inclusion of the vertex y in C is by contractions.

Definition 2.70. Let $\{x, y\} \in E(G)$. The contraction of $\{x, y\}$ is the operation that replaces the two vertices x and y by one new vertex — say xy — whose neighborhood is defined as

$$N(xy) = (N(x) \cup N(y)) \setminus \{x, y\}.$$

The operation that includes y in C could be replaced by contracting the edge $\{R, y\}$.

BACK TO BUSINESS:

Theorem 2.71. *There exists an algorithm — that runs in $O(n^4)$ time — and that computes all minimal clique separators in a graph G.*

Proof. The algorithm we propose computes a feasible partition $\{X, S, C\}$ of $V(G)$. When S is a clique, it recursively computes $\sigma(X)$. All minimal clique separators in $G[X]$, with S tagged on, are minimal clique separators in G.

Next, all edges are added to make S a clique. Denote the subgraph induced by $C \cup S$ as $C \cup \bar{S}$. Recursively, count the minimal clique separators in $C \cup \bar{S}$ that are cliques in G.

The recursions take place on subsets of V that form a partition of $V(G)$, namely X and $C \cup S$. The overhead is dominated by the computation of the feasible partition, which, by (2.41), takes $O(n^3)$ time.

This proves that the time used by this algorithm is $O(n^4)$.

This completes the proof.   □

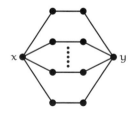

Figure 2.7: This example shows that the number of minimal separators in a graph can be exponential.

Remark 2.72. The number of minimal separators in a graph may be exponential. — For example — consider the graph in Figure 2.7 on the previous page . An $x\,|\,y$ – separator must contain a vertex of each $x\sim y$ – path. If there are t such paths, then there are 2^t minimal $x\,|\,y$ – separators and the graph has $2t + 2$ vertices.

There exists an algorithm that lists all minimal separators in a graph with <u>polynomial delay</u> . That is, the algorithm produces all minimal separators, and it spends polynomial time — either to produce a next one or to end.

Exercise 2.39

Let G be a graph. Consider the following set R of betweenness relations :

$(x, y, z) \in R \quad \Leftrightarrow$

              there is a minimal clique separator that

                      contains y and separates x and z . (2.42)

Design an algorithm that finds a total ordering $\leqslant$ of V satisfying

        $(a, b, c) \in R \quad \Leftrightarrow \quad a < b < c \quad \text{or} \quad c < b < a.$

2.9.3 *Another Intermezzo: Trivially perfect graphs*

When a graph G is connected but has no induced C_4 and no induced P_4 then it has a UNIVERSAL VERTEX (that is a vertex adjacent to all others). This was proved by Wolk in 1961. Let G be a connected graph without induced C_4 or P_4. Then there exists a rooted tree T with $V(T) = V(G)$ such that any two vertices x and y are adjacent in G if and only if one lies on the path to the root of the other one.

Figure 2.8: The figure shows the paw. It has no induced P_4 and no induced C_4. It does have a universal vertex.

The graphs without induced P_4 nor C_4 were given the epithet 'trivially perfect' by Golumbic. The reason for that name is that in any induced subgraph of the graph the independence number equals the number of maximal cliques.

2.10   Vertex ranking

Definition 2.73. Let G be a graph and let $t \in \mathbb{N}$. A t-ranking is a coloring $c : V \to [t]$ which satisfies the following property. For any two vertices x and y with $c(x) = c(y)$ any $x \sim y$-path contains a vertex z with $c(z) > c(x)$.

The rank of G is denoted as $\chi_r(G)$ and it is the smallest t for which G has a t-ranking. — CLEARLY — $\chi \leqslant \chi_r$ since a t-ranking is a proper coloring.

Exercise 2.40

Let G be a connected graph and assume it has a t-ranking. There exists at most one vertex that has color t.

Exercise 2.41

What is the rank of C_4? What is the rank of P_4? What is $\chi_r(G)$ when G is trivially perfect?

Exercise 2.42

Let G be a graph which is not a clique. Show that

$$\chi_r(G) \quad = \quad \min_S \max_C |S| + \chi_r(C)$$

where S varies over the set of minimal separators in G and C varies over the collection of components of $G - S$.

2.10.1   Permutation graphs

Definition 2.74. A graph is a permutation graph if it is the intersection graph of a set of line-segments in the plane that have their endpoints on two parallel lines.

Figure 2.9: The figure shows a permutation model. The graph has the black line-segments as vertices; two being adjacent if the line-segments intersect.

Exercise 2.43

(1) Show that a graph is a permutation graph if and only if its complement is a permutation graph.

(2) Show that a permutation graph is a comparability graph — that is it has a <u>transitive</u> orientation of its edges. [41]

[41] That is, if $x \rightarrow y$ and $y \rightarrow z$ then $\{x, z\} \in E$ and $x \rightarrow z$.

(3) Show that a graph is a permutation graph if and only if it and its complement are comparability graphs.

(4) Design efficient algorithms to compute α and ω for permutation graphs. You may assume that a permutation model is given as input.

(5) Show that any permutation graph has a <u>dominating pair</u> — that is — a pair of vertices x and y with the property that any $x \sim y$-path is a dominating set in the graph. [42]

[42] A set $D \subseteq V$ is a dominating set if every vertex that is not in D has a neighbor in D.

2.10.2   Separators in permutation graphs

CONSIDER A SCANLINE in a permutation model — that is — a new line-segment s with its endpoints on the two parallel lines. The line-segment s splits the set of vertices in three parts:

(a) vertices of which the line-segment lies to the left of s

(b) vertices of which the line-segment lies to the right of s

(c) vertices of which the line-segment crosses with s.

Exercise 2.44

Let G be a permutation graph and let S be a minimal separator in G. In any model for G there exists a scanline s such that the vertices of S are the line-segments that cross s.

COROLLARY: the number of minimal separators in a permutation graph is $O(n^2)$.

Hint: Consider a permutation model. Remove the vertices of S from the model. Let C_1 and C_2 be two components of $G - S$ such that every vertex of S has a neighbor in both. For the scanline s take any line-segment in the model that lies between any two line-segments of C_1 and C_2 and that crosses no other line-segments. That line-segment must exist since C_1 and C_2 form connected parts in the diagram. Since a vertex of S has a neighbor in C_1 and in C_2 it must cross s. The only vertices that were removed from the model are in S so all line-segments that cross s are in S.

Figure 2.10: Illustration of a green scanline.

2.10.3   Vertex ranking of permutation graphs

Definition 2.75. Let s_1 and s_2 be two parallel scanlines in a permutation model. The piece $C(s_1, s_2)$ is the subgraph of G induced by the following sets of vertices.

i. vertices of which the line-segment s between s_1 and s_2

ii. vertices that cross at least one of s_1 and s_2.

Definition 2.76. Let $C(s_1, s_2)$ be a piece. A scanline t <u>splits the piece</u> if t is between s_1 and s_2.

Theorem 2.77. *There exists an* $O(n^6)$ *algorithm to compute the rank of a permutation graph.*

Proof. A permutation model can be constructed in linear time via a modular decomposition. [43]

[43] R. McConnell and J. Spinrad, *Modular decomposition and transitive orientation*, Discrete Mathematics **201** (1999), pp. 189–241.

Organize the pieces according to the number of vertices that are in it. To compute χ_r for a piece $C(s_1, s_2)$ try all scanlines that split the piece into smaller pieces. By Exercise 2.42 the rank of the piece can be computed from the maximal rank of a smaller piece and the size of the separator. $\square$

2.11 Cographs

Definition 2.78. A <u>cograph</u> is a graph without induced P_4.

Figure 2.11: The figure shows the claw, ie, $K_{1,3}$, on the left and P_4 on the right.

ONE IMPORTANT OBSERVATION is that cographs are closed under complementation. — That is — when G is a cograph then so is its complement $\bar{G}$. That is so because

$$P_4 \text{ is isomorphic to its complement } \bar{P}_4.$$

— One other thing — a graph is a cograph if and only if all of its components induce cographs. That is so because an induced P_4 cannot have points in more than one component. Thus, whenever G_1 and G_2 are two cographs, we can create two new cographs by taking their union $G_1 \oplus G_2$ and their join $G_1 \otimes G_2$. [44] Folklore asserts that this property characterizes cographs.

[44] In the union $G_1 \oplus G_2$ the edges are just the edges of G_1 and G_2. In the join $G_1 \otimes G_2$ we add all edges that have one endpoint in G_1 and the other in G_2. So

$$\bar{G}_1 \otimes \bar{G}_2 = \overline{G_1 \oplus G_2}.$$

Theorem 2.79. *A graph is a cograph if and only if every induced subgraph has only one vertex — or else — it or its complement is disconnected.*

By Theorem 2.79 each cograph G is represented by a <u>cotree</u> — that is — a rooted tree T and a bijection from the leaves of T to $V(G)$. The internal nodes of T are labeled: each internal node has a label $\oplus$ or $\otimes$. Two vertices in G are adjacent if and only if their lowest common ancestor in T is labeled with $\otimes$.

Definition 2.80. A pair of vertices x and y in a graph is a <u>twin</u> if every third vertex z is either adjacent to both or neither of x and y — that is — a twin is a module in the graph with two elements. A twin is a true twin if the pair is adjacent and a false twin if the pair is not adjacent.

Theorem 2.81. *A graph is a cograph if and only if every induced subgraph with at least two vertices has a twin.*

Exercise 2.45

Characterize bipartite cographs.

Exercise 2.46

Design an efficient algorithm to compute the rank χ_r of a cograph.

2.11.1 Switching cographs

Definition 2.82. A two-graph is a pair (V, Δ) where V is a finite set and where Δ is a collection of <u>triples</u> in V with the property that every set with 4 elements from V contains an even number of triples that are in Δ. [45]

[45] Two-graphs look a lot like graphs but in two-graphs the 'edges' are triples.

Exercise 2.47

Let G be a graph and let Δ be the 'odd triples in G' — that is — the set of triples in V that carry an odd number of edges between them. Show that this is a two-graph.

Exercise 2.48

Let G be a graph and let $S \subseteq V$. A <u>switch</u> of G with respect to S replaces all edges that have one end in S by nonedges and all nonadjacent pairs with one end in S by edges. Show that a switch does not change the set of odd triples.

CALL TWO GRAPHS G AND H SWITCH-EQUIVALENT if one is obtained from the other via a switch. Van Lint and Seidel showed that two-graphs are switch equivalence - classes.

Exercise 2.49

A two-graph is the switch equivalence - class of a cograph if and only if it does not contain the <u>pentagon</u> — that is — the set of odd triples of the C_5.

The switch - class of cographs is characterized by the following property. There exists a tree T without vertices of degree two. Let V be the set of leaves in T. Since T is bipartite it has a unique coloring with two colors — say black and white. Define Δ as the collection of triples $\{x, y, z\}$ in V if the paths that connect the three meet in a black vertex. Then (V, Δ) is a two-graph and the two-graphs obtained in this manner are exactly the switching - class of cographs.

Exercise 2.50

Show that a graph can be switched to a cograph if and only if it does not contain C_5, the bull, the gem or the cogem as an induced subgraph — that is — no subgraph with 5 vertices switches to C_5.

Show that a graph G can be switched to a cograph if and only if every induced subgraph with at least two vertices has a twin or an anti-twin. [46]

[46] An <u>anti-twin</u> in a graph is a pair of vertices x and y such that every third vertex z is adjacent to exactly one of them.

Exercise 2.51

Design a linear - time algorithm to compute χ for graphs in the switch - class of cographs. You may assume that a tree - model is given as input.

HINT: Show that $\chi = \omega$ for any graph that switches to a cograph.

Exercise 2.52

The rank of the adjacency matrix of a cograph is equal to the number of distinct non - zero rows. [47]

For a proof of this see eg the following paper.

G. Royle, The rank of a cograph, *Electronic Journal on Combinatorics* **10**, (2003), Note 11.

In this paper the question is raised whether any other "natural" classes of graphs satisfy this property. What can you say about the rank of a graph that switches to a cograph?

HINT: P_5 has eigenvalue zero — so — the rank of P_5 is not equal to the number of distinct rows in the adjacency matrix of P_5. — Furthermore — P_5 switches to a cograph.

Switching does not change the eigenvalues of the Seidel - matrix of the graph. The Seidel - matrix has:

1. zeros on the diagonal

2. -1 at entries that represent edges in the graph

3. $+1$ at entries that represent nonedges in the graph.

So it is the matrix $J - I - 2A$ where A is the ordinary adjacency matrix of the graph. When the graph is regular the eigenspaces of A and $J - I - 2A$ are the same.

The spectrum of the Seidel matrix is the same for any two graphs that are switching - equivalent. For a proof see Corollary 3.3 in the paper below.

J. Seidel, *A survey of two - graphs*, Atti Convegno Internazionale Teorie Combinatorie, Tomo 1 (Rome, Italy, September 3-15, 1973), Acdemia Nazionale dei Lincei, Roma (1976), pp. 481–511.

[47] For graphs in general this is an upperbound.

Seidel's favorite graph $C_5 + K_1$ has Seidel spectrum $\{\sqrt{5}^{(3)}, -\sqrt{5}^{(3)}\}$. This graph switches to the net (the complement of the 3-sun).

2.12 Parameterized Algorithms

LET'S START WITH A BIRD'S - EYE VIEW.

Exercise 2.53

Show that there is an algorithm that checks if a graph G has a set $S \subseteq V$ — $|S| \leqslant k$ — such that $G - S$ is a cograph — where $k \in \mathbb{N}$ is a <u>parameter</u>. Your algorithm should run in

$$O\left(4^k \cdot n^3\right).$$

The algorithm is a function of k.

Hint: First design an $O\left(n^3\right)$ algorithm that finds an induced P_4 — if there exists one — eg, using a matrix multiplication. Next, if P is an induced P_4, then branch, each time putting a different point of P in S. Since $|S| \leqslant k$, the depth of the recursion is k.

Unless $P = NP$ no NP - complete problem can be solved in polynomial time. All known algorithms that solve NP-complete problems are exponential. — Therefore — we wish to design 'fast' exponential - time algorithms to solve hard problems.

I've heard that before!

Parameterized algorithmics is a theory designed to help you do this. The genesis of the theory is the notion of a parameterized problem.

Definition 2.83. A parameterized problem is a pair (P, k) where P is a computational problem and k is a parameter that differentiates some solutions from others.

The parameter expresses the 'size' of a solution but the definition of a size is up to the composer of the problem. — Below — we give three examples of parameterized problems.

When the values of the parameter k ranges over $\mathbb{N}$ then a parameterized problem is an ordering of the solutions (by their size).

1. Let I be the <u>vertex cover - problem</u>: Let G be a graph. Find a smallest set $S \subset V(G)$ such that $G - S$ is empty.

    Let $k \in \mathbb{N}$. The parameterized vertex cover - problem (I, k) is to find a vertex cover S with $|S| \leqslant k$.

$G - S$ is empty if

$$E(G - S) = \varnothing$$

2. Let II be the edge domination - problem: Let G be a graph. An
   edge dominating set is a dominating set in the linegraph $L(G)$
   — in other words — it is a set $M \subseteq E(G)$ of edges in G such
   that $E(G - V(M)) = \varnothing$. The problem asks for a smallest edge
   dominating set.

   The parameterized problem edge domination - problem (II, k) is
   to find an edge dominating set with at most k edges.

For a set M of edges let
$V(M)$ be the set of end-
points of edges in M — that
is — $V(M) = \cup_{e \in M} e$.

3. Problem III is the feedback vertex set - problem. This problem
   applies to graphs G that may have loops and multiple edges. The
   problem is to find a smallest set $S \subset V(G)$ such that $G - S$ has
   no cycles.

   The parameterized problem parameterized feedback vertex set
   problem (III, k) is to find a feedback vertex set of size at most k.

Definition 2.84. A fixed parameter - algorithm is an algorithm
that solves a parameterized problem (P, k) with parameter $k \in \mathbb{N}$
in

$$O (f(k) \cdot |P|^c)$$

time. Here $f : \mathbb{N} \to \mathbb{N}$ is a computable function and $c \in \mathbb{N}$.

For ease of discussion in Definition 2.84 we assume that
the parameter k ranges over
the natural numbers.

A computable function is a
function which can be evaluated via an algorithm.

Notice that c is a constant; (not a parameter); the run - time
of a fixed parameter algorithm is a polynomial in the size of the
instance of problem P. The influence of the parameter k on the
run - time is some arbitrary function $f(k)$. The algorithm runs
in polynomial - time if we let the parameter k be a constant (then
$f(k)$ disappears in the Big-Oh).

A constant is a natural number.

When P is NP - complete the
function f is not a polynomial.

There are problems that can not be solved by a fixed parameter -
algorithm. The 'good guys' are called fixed parameter - tractable.

Definition 2.85. Let (P, k) be a parameterized problem. The
problem is fixed parameter tractable (FPT) if there is a fixed
parameter - algorithm that solves (P, k).

Nobody knows of an algo-
rithm that solves the pa-
rameterized problem $\omega \geqslant k$
and that runs in time $O(n^c)$
where c does not depend on
k. The clique problem is
$W[1]$ - hard.

Remark 2.86. In their book Downey and Fellows introduce a W-hierarchy to capture the hardness of parameterized problems. In this hierarchy $W[0] = \mathsf{FPT}$. A parameterized problems is $W[1]$ - hard if there is no fixed - parameter algorithm to solve it (under certain logical assumptions).

This section has too many imprecise definitions and — unexplained — assumptions. Let's hope that — in practice — it all works out fine.

2.13 The bounded search technique

LET'S TAKE A LOOK AT A BASIC TECHNIQUE to design fixed parameter - algorithms. The bounded search technique is best explained by example.

2.13.1 Vertex cover

Our example is the vertex cover problem parameterized by the size of the solution. Let G be a graph and let S be a vertex cover in G — that is

We want a vertex cover with at most k vertices.

$$e \in E(G) \quad \Rightarrow \quad e \cap S \neq \varnothing.$$

The following algorithm searches for a vertex cover of size $\leqslant k$ (where the parameter k ranges over $\mathbb{N} \cup \{0\}$).

If the graph has no edges then $\varnothing$ is a solution. If $k = 0$ and $E(G) \neq \varnothing$ then there is no solution. Otherwise pick an edge e from the graph and build a search tree with that edge as a root.

Every edge of the graph has at least one endpoint that is in a vertex cover. The search tree tries both endpoints and searches for a small vertex cover that contains an endpoint.

The search tree (with root e) branches at the root into two subtrees; each subtree is labeled with an endpoint of e. The subtrees are evaluated as follows.

The selected endpoint of e is put in S and this endpoint is deleted from the graph. The parameter k decreases with 1 and the subtree searches for a vertex cover of size $k - 1$ (in the remaining graph).

The size of the search tree is $O(2^k)$ (since the depth of the search tree is at most k and every node has at most two children). This proves the following theorem.

Theorem 2.87. *The vertex cover problem is fixed-parameter tractable and can be solved in* $2^k \cdot |I|^{O(1)}$ *time.*

2.13.2 Edge dominating set

The parameterized vertex cover problem is easy to solve via the building of a search tree of size 2^k.

A DIFFERENT KETTLE OF FISH is the parameterized edge dominating set problem. To build a search tree for this problem we would want to find a set of edges M which satisfies the following two conditions.

1. |M| is bounded by some function of k

2. any solution to the parameterized edge domination problem has at least one edge in M.

This road doesn't look very appealing. We take a different approach.

Instead of trying to locate the edges of a solution we first find a collection of suitable sets $S \subseteq V(G)$ that are endpoints of a solution. Step two is to find a a solution — that is — a set of edges that solves the parameterized edge domination problem and that contains all elements of S.

Why doesn't this idea work? What happens to the problem if we assume that the degree of the graph is at most 3?

Exercise 2.54

Let M be a minimum edge dominating set. Then V(M) is a vertex cover.

Exercise 2.55

Prove or disprove:
Let S be a minimal vertex cover. A minimum edge dominating set M which satisfies $S \subseteq V(M)$ can be computed in polynomial time

as follows. Initialize M as a maximum matching in $G[S]$. For each vertex $x \in S \setminus V(M)$ add one edge of G that contains x to M.

The exercises show that our job is done if we can find the right minimal vertex cover.

Theorem 2.88. *The parameterized edge dominating set problem is fixed-parameter tractable and can be solved in $4^k \cdot |II|^{O(1)}$ time.*

Proof. The following algorithm solves the parameterized edge domination problem.

1. generate the set S of all minimal vertex covers with at most $2k$ elements

2. for each $S \in S$ compute a minimum edge dominating set M with $S \subseteq V(M)$ as in Exercise 2.55

3. Output M when $|M| \leqslant k$.

This proves the theorem. $\square$

2.13.3 Feedback vertex set

In the previous two examples we made use of the fact that a small 'local' part of the graph contain a solution. The feedback vertex problem lacks this property.

IT IS ALWAYS A GOOD IDEA to reduce the graph before going on a venture that takes exponential time — so — that's what we do.

1. if a vertex is not in any cycle then we delete the vertex from the graph

2. if there are two vertices with more than two edges running between them then we delete one of those edges

A solution is some 'small' set of vertices that hits all cycles of the graph. What happens to the feedback vertex set problem if we assume that the graph has no induced cycles of length more than 3?

Exercise: Show that the reductions take polynomial time and that they are 'safe:' the graph has a solution if and only if the reduced graph has a solution.

3. if a vertex is in a loop then we delete the vertex from the graph and we decrease the parameter k by 1 (the vertex is in every solution)

4. if x has only one neighbor y and if there are at least two edges between x and y then we remove x and we add a loop at y

5. if a vertex is in exactly two edges that connect it to two different neighbors then we remove the vertex and replace it with an edge that connects the two neighbors.

Definition 2.89. A graph is reduced if none of the operations above apply.

> A reduced graph has no loops and at most two parallel edges between any two vertices. Furthermore every vertex is in at least three edges.

We turn our attention to the reduced graph. A solution is some 'small' set X of vertices that hits all cycles. If we remove X from the graph the remainder is a 'large' forest F. All the vertices of F that have at most two neighbors in F must have neighbors in X. Not every vertex of X has a small degree; that is so because there is no large forest with only a few leaves and only a few vertices of degree two.

LET'S WORK THIS OUT. Order the vertices in the graph in a descending order of their degree say

$$d(v_1) \geqslant \cdots \geqslant d(v_n).$$

Define the set of high - degree vertices as the set of the first $\lceil 3k/2 \rceil$ vertices in this order:

$$H = \{v_1, \ldots, v_{\lceil 3k/2 \rceil}\}.$$

Call the elements of H the vertices of high degree.

Lemma 2.90. *Let* $G = (V, E)$ *be a reduced graph. Any feedback vertex set in* G *of size at most* k *contains at least one vertex of high degree.*

Proof. Let X be a feedback vertex set and let $F = V \setminus X$. Then $G[F]$ is a forest and the number of edges in $G[F]$ is at most $|F| - 1$. We have that

$$\sum_{z \in X} d(z) \;\geqslant\; |E| - |F| + 1.$$

Assume that there is a feedback vertex set X of size at most k and assume that $X \cap H = \varnothing$. We show that this leads to a contradiction.

Following the idea outlined above we concentrate on the number of edges that run between F and X.

Let $f = |F \setminus H|$. Each vertex in $F \setminus H$ has degree in G at least three since G is reduced. Let $a = d(v_{\lceil 3k/2 \rceil})$. Then $a \geqslant 3$.

By assumption $H \subseteq F$. We have that

$$\sum_{z \in F} d(z) \;\geqslant\; \lceil 3k/2 \rceil \cdot a + 3 \cdot f.$$

The induced graph $G[F]$ is a forest and the number of edges in $G[F]$ is at most $|F| - 1$. Thus, the number of edges that run between F and X is at least

$$\lceil 3k/2 \rceil a + 3f - |F| + 1 \;=\; \lceil 3k/2 \rceil (a - 1) + 2f + 1.$$

ON THE OTHER HAND each vertex in X has degree at most a — and so — the number of edges between X and F is at most $k \cdot a$.

So we have that

$$\lceil 3k/2 \rceil (a - 1) + 2f + 1 \;\leqslant\; k \cdot a.$$

and this leads to $a < 3$. This contradicts the fact that each vertex in G has degree at least 3. $\qquad\square$

Lemma 2.90 allows the following parameterized algorithm for feedback vertex set.

Reduce the graph. The vertices of high degree in the reduced graph form the root of a branching. A selected vertex is put in the solution set and deleted from the graph.

Each branching operation results in at most $\lceil 3k/2 \rceil$ subproblems. The depth of the recursion is at most k. This proves the following theorem.

Exercise: Check all this.

Theorem 2.91. *The parameterized feedback vertex set problem is fixed-parameter tractable and can be solved in* $(1.5k)^k \cdot |\text{III}|^{O(1)}$ *time.*

Exercise 2.56

A set of vertices in a graph is called a $\underline{P_3 \text{ - cover}}$ if each path of length two in the graph contains at least one vertex from the set.

Please design a fixed parameter algorithm for the problem to decide if a graph has a P_3 - cover of size at most k (where k is a parameter).

Hint: Any P_3 - cover contains at least one of the three vertices of any path of two edges.

2.13.4 Further reading

The fastest parameterized algorithm for the vertex cover problem runs in $1.2738^k n^{O(1)}$ time. This was obtained by Chen, Kanj and Xia in 2010.

The edge dominating set problem can be solved in $2.3147^k n^{O(1)}$ time by Xiao, Kloks and Poon. This is further improved to $2.2351^k n^{O(1)}$ by Iwaide and Nagamochi.

For the feedback vertex set problem, there is a $2.7^k n^{O(1)}$-time randomized algorithm by Li and Nederlof. There is also a $3.46^k n^{O(1)}$-time deterministic algorithm by Iwata and Kobayashi.

References

R. Downey and M. Fellows, *Fixed Parameter Tractability and Completeness*. Dagstuhl workshop Complexity theory: current research, 1992, pp. 191 – 225.

J. Chen, I. Kanj and G. Xia, *Improved upper bounds for vertex cover*, Theor. Comput. Sci. 411 (2010), pp. 3736 – 3756.

M. Xiao, T. Kloks and S. Poon, *New parameterized algorithms for the edge dominating set problem*, Theor. Comput. Sci. 511 (2013), pp. 147 – 158.

K. Iwaide and H. Nagamochi, *An Improved Algorithm for Parameterized Edge Dominating Set Problem*, J. Graph Algorithms Appl. 20 (2016), pp. 23 – 58.

J. Li and J. Nederlof, *Detecting Feedback Vertex Sets of Size* k *in* $O*(2.7^k)$ *Time*, Proceedings of SODA 2020, pp. 971 – 989.

Y. Iwata and Y. Kobayashi, *Improved Analysis of Highest-Degree Branching for Feedback Vertex Set*, Algorithmica 83(8) (2021), pp. 2503 – 2520.

2.14 Matchings

A matching in a graph G is a set of edges of which no pair shares an endpoint.

Figure 2.12: The 5-wheel

Definition 2.92. Let G be a graph with at least one edge. A set

$$S \subseteq E$$

is a <u>matching</u> if S is an independent set in $L(G)$.

We denote a matching of maximal cardinality in G by

$$\nu(G) = \alpha(L(G)).$$

A matching of maximal cardinality is called a **maximum** matching.

Exercise 2.57

Check Definition 2.92 with the text above it.

ν is the 13th Greek alphabet letter 'nu.' In the English alphabet, the 13th letter is m, for matching.

2.15 Independent Set in Claw - Free Graphs

Figure 2.13: The claw

CLAW - FREE GRAPHS GENERALIZE LINEGRAPHS. Of course they can be recognized in polynomial time. Minty designed a polynomial–time algorithm to find maximum independent sets in claw–free graphs. We describe this algorithm. Since linegraphs are claw–free, this implies that there is a polynomial–time algorithm to find a maximum matching in a graph.

When $\mathcal{F}$ is some finite collection of graphs, then the class of $\mathcal{F}$-free graphs is the set of those graphs that have no induced subgraph isomorphic to an element of $\mathcal{F}$.

BTW Harary exhibits a complete list of nine forbidden induced subgraphs that characterize linegraphs. (For example — the 5-wheel is claw-free but not a linegraph).

2.15.1 The Blossom Algorithm

Let us first recapitulate Edmonds' algorithm to compute a maximum matching in a graph.

A chain is a sequence of vertices

$$P = [v_0 \quad \cdots \quad v_t]$$

such that successive elements are adjacent. Let M be a maximal matching. Let X be the set of vertices that are not in any line of M.

Berge showed that M is not maximal if and only if there exists an M-augmenting path — that is — a path that starts and ends with distinct points in X and whose edges alternate between M and $E \setminus M$. Berge's original proposal to find an augmenting path via a depth-first-search procedure did not work because the path could end up in an odd cycle. It was Edmonds' idea to shrink the cycle into one new point and start the search afresh.

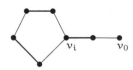

Figure 2.14: The figure shows a flower. The thick lines represent edges of M.

Definition 2.93. A chain $P = [v_0 \quad \cdots \quad v_t]$ is alternating if for each i exactly one of the edges $\{v_{i-1}, v_i\}$ and $\{v_i, v_{i+1}\}$ is in M.

We first show that we can find a shortest alternating chain with endpoints in X. Let A be the set of edges of a directed graph (V, A) defined as follows

$$A = \{u \to v \mid \exists_{x \in V} \ \{u, x\} \in E \ \text{ and } \ \{x, v\} \in M\}.$$

Then each alternating chain which starts and ends in X is a directed path from X to $N_G(X)$ in (V, A). Thus we can find an alternating chain in G in polynomial time.

Definition 2.94. An alternating chain

$$P = [v_0 \quad \cdots \quad v_t]$$

is a <u>flower</u> if

1. t is odd

2. $v_0, \ldots, v_{t-1}$ are distinct

3. $v_t = v_i$ for some $i < t$ where i is even.

The circuit $[v_i, \ldots, v_t]$ is called the <u>blossom</u> of the flower.

Lemma 2.95. *A shortest alternating chain*

$$P = [v_0 \quad \cdots \quad v_t]$$

from X to X is either an augmenting path or $[v_0 \quad \cdots \quad v_j]$ is a flower for some $j \leqslant t$.

Proof. Assume P is not a path. Let $i < j$ be such that $v_i = v_j$ and such that j is as small as possible. Thus $v_0, \ldots, v_{j-1}$ are distinct.

If $j - i$ is even delete $v_{i+1}, \ldots, v_j$ from P and obtain a shorter alternating chain from X to X.

Assume $j - i$ is odd. The case where j is even and i is odd cannot occur since this would imply that two edges of the matching meet in v_i.

Thus we may assume that j is odd and i is even which implies that

$$[v_0 \quad \cdots \quad v_j]$$

is a flower. $\square$

Let B the the blossom of a flower. The graph G/B replaces the set of vertices of B by one new vertex. We call this new vertex B. The edges of the graph G/B are

$$E(G/B) = \{\{x,y\} \mid \{x,y\} \in E \quad \text{and} \quad x,y \notin B\} \cup$$
$$\{\{B, y\} \mid y \in N_G(B)\}. \quad (2.43)$$

For the matching M in G we let M/B be the set of corresponding edges in G/B. An edge of M with both ends in B is not an edge of M/B.

Lemma 2.96. *Let*

$$B = [v_i \quad \cdots \quad v_t]$$

be a blossom in G. *Then* M *is a matching of maximal cardinality in the graph* G *if and only if* M/B *is a matching of maximal cardinality in the graph* G/B.

Proof. Assume that M/B is not maximum in G/B. Let P be an augmenting path. If P does not contain B then it is an augmenting path in G.

Assume that P enters B by an edge

$$\{u, B\} \notin M/B.$$

Thus $\{u, v_j\} \in E$ for some $j \in \{i, \ldots, t\}$. If j is odd then replace B in P by $[v_j \, v_{j+1} \quad \cdots \quad v_t]$. If j is even then replace B by $[v_j \, v_{j-1} \quad \cdots \quad v_i]$. In both cases we obtain an augmenting path in G.

Now assume that $|M|$ is not maximal. We may assume that $i = 0$ — that is — $v_i \in X$. Otherwise we can replace M by the symmetric difference $M \div Q$ where Q is the set of lines in the chain $[v_0 \quad \cdots \quad v_i]$.

Let $P = [u_0 \quad \cdots \quad u_s]$ be an augmenting path in G. When P does not visit B then P is an augmenting path in G/B. If P visits B then we may assume that $u_0 \notin B$ since otherwise we can replace P by its reverse. Let u_j be the first vertex of P in B. Then $[u_0 \quad \cdots \quad u_{j-1} \, B]$ is an augmenting path in G/B. — Thus — $|M/B|$ is not maximal. $\qquad \square$

Edmonds' algorithm can be implemented to run in $O(n^2 \cdot m)$ time. Micali and Vazirani show that a maximum matching can be computed in $O(\sqrt{n} \cdot m)$ time.

2.15.2 Minty's Algorithm

We assume that the reader is familar with Edmonds' algorithm to find a maximum matching in graphs. Minty's algorithm computes $\alpha(G)$ in claw-free graphs by reducing it to that of finding a maximum matching in an auxiliary graph constructed from the input graph.

Let G be claw-free and let B be a maximal independent set in G. Color the vertices of B black and the others white. Notice that every white vertex has 1 or 2 black neighbors.

An augmenting path is a path that runs between two white vertices — that each have only one black neighbor — and of which the white vertices form an independent set.

Lemma 2.97. *When two white vertices — both having two black neighbors — are adjacent then they have a common black neighbor.*

Proof. Let x and y be two white vertices that are adjacent. Notice that

$$| \, (N(x) \cup N(y)) \cap B \, | \geqslant 4$$

implies a claw with one of the two whites as a center (since the black vertices form an independent set). This is a contradiction. □

If there exists an independent set of cardinality larger than $|B|$ then there must exist an augmenting path. Minty's algorithm finds an augmenting path — if it exists — as follows.

Definition 2.98. A <u>wing</u> is a nonempty subset of white vertices that is a single white vertex that has only one black neighbor or the common neighborhood of two black vertices.

We refer to the wings of the second kind as the 'tipped wings.'

Definition 2.99. A black vertex is <u>regular</u> if either

1. it is incident with a white vertex that has only one black neighbor

2. it is incident with at least three tipped wings.

Lemma 2.100. *Let* b *be a regular black vertex of the second kind. Let* $p, q, r \subset N(b)$ *and let them be in different tipped wings incident with* b. *Then the number of pairs in* $\{p, q, r\}$ *that are edges is odd.*

Proof. The graph would have a claw when the number of edges in $\{p, q, r\}$ were 0 or 2 (either with b as a center or the one of p, q and r that is adjacent to the other two). $\square$

2.15.3 A Cute Lemma

In an attempt to find an augmenting path we may try to find an augmenting path that runs between a fixed pair of white vertices that have one black neighbor. By trying all $O(n^2)$ feasible pairs this will find an augmenting path if it exists.

This approach has the advantage that we can reduce the graph: Let s and t be two white vertices that are not adjacent and that have each exactly one black neighbor. In the search for an augmenting s ~ t-path we can remove all white neighbors of s and t and all other white vertices that have only one black neighbor. We refer to this graph as the reduced structure.

Minty first shows that the neighborhood of any regular back vertex has a partition into two classes such that all nonedges run between vertices in different classes. For a regular vertex b that is adjacent to s or to t one of the two classes is s or t and the other class is $N(b) \setminus \{s, t\}$. The following lemma concerns itself with the regular vertices that are incident with at least three tipped wings.

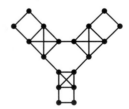

Figure 2.15: A claw-free graph; this one is a domino: every vertex is in two maximal cliques.

Lemma 2.101. *Let* b *be a black vertex that is incident with at least three tipped wings. There exists a unique partition of* $N(b)$ *into at most two parts such that all nonedges that run between whites in different wings have their endpoints in different parts and all edges between whites in different wings have their endpoints in similar parts.*

Proof. Notice that if x, y and z are three white vertices in different tipped wings at b then the number of edges among $\{x, y, z\}$ is odd. We call a triple $\{x, y, z\}$ with an odd number of edges among them an <u>odd triple</u>.

First notice that — when each triple in a graph is odd — then the graph is a union of at most two cliques. In that case we are done.

Consider a wing S and let $z \notin S$. Define a binary relation L_z on S as follows. Any two elements x and y in S are related if and only if $\{x, z\}$ and $\{y, z\}$ are both edges or both nonedges.

It is not difficult to check that any $z' \notin S$ produces the same relation $L_{z'} = L_z$ on S — that is — we can write L instead of L_z.

Make two vertices in a wing S adjacent when they are related in L and nonadjacent when they are not related in L. [48] Then every triple in the wing is an odd triple.

It is now easy to check that any triple $\{x, y, z\}$ in $N(b)$ is odd. It follows that $N(b)$ has a partition as claimed in the lemma. □

[48] Notice that any augmenting path can use only one white vertex in a wing. So the graph on the white vertices within a wing is of no importance. The lemma makes no claim on the edges that run between whites that are in the same wing.

Exercise 2.58

Let G be a graph and let Ω be its collection of odd triples. A <u>Seidel switch</u> with respect to some set $S \subseteq V$ changes all edges with one endpoint in S into nonedges and it changes all nonedges with one endpoint in S into edges.

Prove that a Seidel switch does not change Ω. When is a collection of triples the set of odd triples of a graph?

Hint: A collection of triples is the collection of odd triples in a graph if and only if every four elements contain an even number of even triples.

Figure 2.16: Three wings (edges with white endpoints are not shown)

2.15.4 Edmonds' Graph

Notice that we lack a partition — into nonadjacent classes — of the neighborhoods of black vertices whose neighborhoods consists of two tipped wings. Minty calls these vertices irregular.

As a <u>subroutine</u> the algorithm determines whether two regular black vertices — say a and b — are connected by an alternating path whose white vertices form an independent set.

Consider a sequence $(W_0, b_0, W_1, b_1, \ldots, W_n, b_n, W_{n+1})$ where the b_i are irregular black vertices and W_i is the wing at b_i with tip b_{i+1}. We may assume that this sequence is maximal — that is — W_0 and W_{n+1} are wings incident with regular vertices. By dynamic programming we can determine whether there exists a path

$$[a \quad w_0 \quad b_0 \quad w_1 \quad b_1 \quad \cdots \quad w_{n+1} \quad b]$$

such that

Figure 2.17: This is the sun. The sun is claw-free. It is the linegraph of the net.

1. a and b are the tips of W_0 and W_{n+1}

2. a is not adjacent to $w_0 \in W_0$ and b is not adjacent to $w_{n+1} \in W_{n+1}$

3. consecutive white vertices $w_i \in W_i$ and $w_{i+1} \in W_{i+1}$ are nonadjacent.

We refer to such an $a \sim b$-path as an irregular path.

Let N be the number of regular vertices in the graph. Edmonds' graph consists of a matching with N edges, representing the regular vertices. The endpoints of an edge represent the two classes of the partition of the regular black vertex. The graph has two more vertices s and t. These are joined by an edge to the node-classes of the two unique regular black neighbors.

Figure 2.18: This graph is called the <u>bull</u>. The bull is claw-free, but has an edge-contraction that produces a claw. Thus, the class of claw-free graphs, is not closed under edge-contractions. BTW, what is the complement of the bull?

By the subroutine described above for any two classes of regular black vertices we can decide whether they are connected by an augmenting path that uses no irregular vertices. Between any two of the $2N$ endpoints of edges in the matching add an edge if the two regular black vertices are connected by an irregular path that uses the two classes.

Lemma 2.102. *There exists an augmenting $s \sim t$-path if and only if Edmonds' graph has an augmenting path.*

We leave it as an exercise to check the correctness.

Theorem 2.103. *There exists an* $O(n^5)$ *algorithm that computes a maximum independent set in claw-free graphs.*

Proof. The problem reduces to finding a augmenting path in Edmonds' graph. The existence of irregular paths can be computed in (overall) $O(n^2)$ time. Finding an augmenting path in Edmonds' graph can be done in $O(n^3)$ time via the blossom algorithm. Since the algorithm tries all feasible pairs s and t as endpoints of an augmenting path the algorithm runs in $O(n^5)$ time.   □

Exercise 2.59

The problem to find $\alpha(G)$ in a triangle–free graph is NP- —complete. Reduce this problem to the clique problem in claw–free graphs. Thus, finding $\omega(G)$ in claw–free graphs is NP–complete.

Faenza et al. show that the independence number in claw-free graphs can be computed in $O(n^3)$ time.

2.16 Dominoes

A natural generalization of the class of linegraphs of bipartite graphs, is the class of dominoes.

Definition 2.104. A graph is a <u>domino</u> if every vertex is in at most two maximal cliques.

Notice that linegraphs of bipartite graphs are dominoes. [49] Not all linegraphs are dominoes — for example — the linegraph of the diamond is the 4-wheel W_4 and W_4 is not a domino.

[49] I spell: "one domino" and "two dominoes."

Exercise 2.60

Show that every domino has at most n maximal cliques.

Figure 2.19: The linegraph of the diamond, on the left, is the 4-wheel, on the right. The 4-wheel is not a domino.

Exercise 2.61

Show that the class of dominoes is <u>hereditary</u> — that is — a graph G is a domino if and only if every induced subgraph of G is a domino.

Dominoes can be characterized in various ways.

Theorem 2.105. *The following propositions are equivalent.*

1. G *is a domino*

2. G *is* {W_4, claw, gem}*-free.*

Figure 2.20: The gem

Remark 2.106. The class of dominoes can be <u>recognized</u> in linear time. — That is — there is a linear–time algorithm that checks whether a graph is a domino. The algorithms operates by <u>identifying</u> vertices that have the same closed neighborhood. The graph on the equivalence classes is a linegraph (with some additional properties). — Actually — a graph is a domino if and only if its representative is the linegraph of a triangle-free graph in which every vertex is adjacent to at most one pendant vertex.

Exercise 2.62

Let G be a graph. Call two vertices <u>equivalent</u> if they have the same closed neighborhood. Show that this defines an equivalence relation on V(G). The <u>representative</u> R(G) of a graph G is the graph with

$$V(R) = \{\ X\ |\ X \subseteq V(G)\ \text{is an equivalence class}\ \}.$$

Two vertices of R are adjacent if a pair of elements of the classes are adjacent. Design a linear–time algorithm that computes the representative of a graph.

A binary relation is an equivalence relation if it is reflexive, symmetric, and transitive. To be precise, a binary relation ~ is an equivalence relation if ~ satisfies

1. $\forall_x\ x \sim x$,

2. $\forall_x \forall_y\ x \sim y \Rightarrow y \sim x$,

3. $\forall_x \forall_y \forall_z\ (x \sim y$ and $y \sim z) \Rightarrow x \sim z$.

The sets of mutually equivalent vertices are called <u>equivalence classes</u>.

Exercise 2.63

A graph is <u>strongly regular</u> if there are numbers k, λ and μ such that

(i) all vertices have the same degree — that is — the graph is regular with degree k:

$$\forall_{x \in V} \quad d(x) = |N(x)| = k$$

(ii)

$$\forall_{x \in V} \; \forall_{y \in V} \; x \neq y \;\; \Rightarrow$$

$$|N(x) \cap N(y)| = \begin{cases} \lambda & \text{if } \{x, y\} \in E \\ \mu & \text{if } \{x, y\} \notin E. \end{cases}$$

— For example — the Petersen graph is strongly regular with parameters

$$(n, k, \lambda, \mu) = (10, 3, 0, 1).$$

(I) Prove that $L(K_n)$ is strongly regular. Is it a domino?

(II) Prove that the adjacancey matrix of a strongly regular graph satisfies

$$A^2 = k \cdot I + \lambda \cdot A + \mu \cdot (J - I - A)$$

where I is the identity matrix and J is the matrix with all elements equal to 1.

HINT: Notice that $J - I - A$ is the adjacency matrix of $\bar{G}$.

2.17 Triangle partition of planar graphs

Definition 2.107. A graph has a <u>partition</u> if its set of edges can be partitioned into triangles. [50] [51]

In this chapter we show that there is a linear - time algorithm to check if a planar graph has a partition.

[50] A triangle is a clique in the graph with 3 vertices.

[51] Finding a minimum set of triangles that COVERS the edges of a planar graph is NP-complete.

Definition 2.108. Let G be a plane graph — that is — the graph G is planar and is given with an embedding in the plane. A triangle T partitions the plane in two open regions say 'inside' and 'outside.' If both regions contain vertices of G the triangle is <u>separating</u>.

Definition 2.109. Let T be a separating triangle and let $x \in V(T)$. The <u>inside degree</u> of x — say $d(x)$ — is the number of edges that is incident with x and some vertex inside T.
A separating triangle is <u>even</u> if $d(x)$ is even for every $x \in V(T)$.

The dual

Exercise 2.64

Assume that G has an edge which is in only one triangle. Say the triangle has edges e_1, e_2 and e_3. Then G has a partition if and only if $G - \{e_1, e_2, e_3\}$ has a partition.

Exercise 2.65

Show that there is a linear-time reduction to the case where the graph G is biconnected — that is — henceforth we assume that G is biconnected. Furthermore we may assume that every vertex of G has at least <u>three</u> neighbors. [52]

[52] A graph is biconnected if each minimal separator has at least two vertices.

Lemma 2.110. *Let H be the dual of G. If G has a partition then H is bipartite.*

Proof. Assume G has a partition. Let C be a cycle in H. The set of edges of C is a <u>cut</u> in G. Every triangle of the partition has either all its vertices on the same side of the cut or one vertex on one side and two on the other side. This shows that the cut has an even number of edges — and so C is even. □

The triangle partition algorithm

— By now — we may assume the following. [53]   [53] Check !

1. G is biconnected

2. the dual H is bipartite

3. every vertex of G has even degree at least 4

4. every edge of G is in at least two triangles.

Graphs without separating triangles

If the graph has no separating triangles then every triangle is a face.

Lemma 2.111. *Assume* G *has no separating triangle. Then* G *has a partition if and only if every vertex of one color class of* H *has degree 3.*

Proof. Let H_1 be a color class of H and assume that all vertices of H_1 have degree 3. Then the vertices of H_1 form a partition of $E(G)$ into triangles.

Assume that G has a partition. The triangles of the partition are faces and the corresponding vertices in H have degree 3. Between any two of them the distance is even so they form a color class of H.   □

Graphs with separating triangles

Let $\mathcal{P}$ be a partition of $E(G)$ into triangles. A separating triangle $S = \{x, y, z\}$ is of one of the following types.

Type 1. $S \in \mathcal{P}$ or the three edges of S are in triangles with the third vertex inside S

Type 2. the three edges of S are in triangles of $\mathcal{P}$ with the third vertex outside S

e 3. some edge of S is in a triangle of $\mathcal{P}$ with the third vertex inside S and some edge of S is in a triangle of $\mathcal{P}$ with the third vertex outside S.

Exercise 2.66

If a separating triangle S is even then it is of Type 1 or Type 2 in any partition.

HINT: Let S be a separating triangle and let $\mathcal{P}$ ve a partition of the edges of G into triangles. Let G' be the graph induced by S and the vertices inside S. When $S \in \mathcal{P}$ then S is even (otherwise G' has no partition).

Let $\{x, y\} \in E(S)$ and assume that $\{x, y\}$ is in a triangle of $\mathcal{P}$ with the third vertex outside S. Assume that the two other edges of S are in triangle with their third vertex inside S. (So S is of Type 3.)

Remove the edge $\{x, y\}$ from the graph G'. There is a partition of the edges of $G' - \{x, y\}$ into triangles — and so — the degree of x and y must be even in $G' - \{x, y\}$. But then S is not an even triangle in G.

Exercise 2.67

All even separating triangles can be found in linear time.

HINT: Use Baker's method to partition $V(G)$ into layers. (See Theorem 4.246 on Page 323.) [54]

[54] Jiawei Gao, Ton Kloks and Sheung-Hung Poon, *Triangle - partitioning edges of planar graphs, toroidal graphs, and k-planar graphs*, Springer - Verlag, Lecture Notes in Computer Science 7748 (2013), pp. 194–205.

Definition 2.112. A separating triangle is <u>outermost</u> if none of its vertices is inside any other separating triangle.

Consider the graph G^* obtained from G by removing the interior of all outermost even separating triangles. So the graph G^* has no even separating triangles. A <u>special region</u> of G^* is a face that is an outermost even separating triangle in G.

Let $\mathcal{P}$ be a partition of G^*. Every even region or face is of one of two types.

Type a. the special region or face is a triangle of $\mathcal{P}$

Type b. the edges of the boundary are in triangles of $\mathcal{P}$ that have their
third vertex outside.

Lemma 2.113. *Assume* G^* *has a partition. Let* H_1 *and* H_2 *be the two color classes of the dual. Then all vertices of* H_1 *are of Type a and all vertices of* H_2 *are of Type b or vice versa.*

Proof. Along any path in the dual of G^* the types of the vertices must alternate. The dual is connected so all vertices of one color class are of the same type.

This proves the lemma. □

Theorem 2.114. *There exists a linear - time algorithm to find a partition of the edges of a planar graph in triangles.*

Proof. There are only two ways to partition G^*.

Let S be an outermost even separating triangle. If S is labeled Type a then a recursive step checks if the inside including S has a partition. If S is labeled as Type b then the interior (without S) is processed in a recursive step.

A list of all even separating triangles can be found in linear time. All recursive steps are performed on separate subgraphs of G. This proves that the algorithm runs in linear time.

This proves the theorem. □

2.17.1 Intermezzo: PQ - trees

Booth and Luecker introduced PQ - trees in 1976 as a data - structure that is useful for the recognition of eg planar graphs and interval graphs.

Let V be a finite set. A PQ - tree is a rooted tree T and a bijection from the elements of V to the leaves of T. Each internal node is labeled P or Q and has at least two children.

The PQ - tree represents a set of permutations of V:

Implementations of PQ - trees allow linear - time recognition algorithms for interval graphs and for planar graphs.

1. the children of a P - node can be re-ordered in any way

2. the order of the children of a Q - node can be reversed (and that is the only other valid order).

Let G be an <u>interval graph</u> — that is — there is an ordering of the maximal cliques in G

$$C_1 \quad \cdots \quad C_t$$

such that each vertex is contained in a consecutive subset.

Consider the $(0, 1)$ - matrix with V as its columns and the maximal cliques in G as its rows and that has a 1 precisely when a vertex is in a clique. Then the rows can be permuted so that all the ones in a column are consecutive.

Exercise 2.68

Design an algorithm:

INPUT: A graph G and the set of all the maximal cliques in G — say — $\{C_1, \cdots, C_t\}$.

OUTPUT: A permutation of the cliques such that the (clique - vertex) - incidence matrix has all ones in each column consecutive.

HINT: Build a PQ - tree. Start with a tree that has all its leaves adjacent to the root and identified with $C_1, \cdots C_t$. Label the root as a P - node. Add the vertices one by one and rebuild the tree to satisfy the consecutive ones property.

2.18 Games

IT'S ALMOST CHRISTMAS — LET'S PLAY SOME GAMES!

One of my teachers used to say: "Only games played between two people are interesting. Larger groups of players give rise to fights and 1-player games are boring!"

2.18.1 Snake

SNAKE IS A THRILLING GAME played on a graph as follows. —
Of course — the game is played between two players. The 'game
- board' is a graph G. The two players take turns in making (legal)
moves. The player who can't make a legal move loses the game.

The two players together build a path in the graph. One endpoint
of the path is the 'head' and the only legal moves that are available
to a player (when it is his turn) extend the path with one vertex
that is adjacent to the head. [55] A newly added vertex becomes the
head of the path.

Player 1 makes the first move and he chooses one point in the
graph to start the path.

Johan Cruijff was once
the greatest football player
in Holland. One of his
proverbs was: "Football is
a game of mistakes. Who-
ever makes the fewest wins!"

Exercise 2.69

Show that player 1 can easily win the game if the graph has an
isolated vertex.

[55] Of course, only legal
moves are allowed. A player
is allowed to make a legal
move when it is his turn.

C. Berge, *Combinatorial
games on a graph*, Discrete
Mathematics **151** (1996),
pp. 59–65.

> We are interested in the question whether player 1 has a winning
> strategy.

Snake was invented by Berge and he formulated the following
beautiful theorem in 1996.

Theorem 2.115. *Player 1 wins the game snake if and only if
the graph has no perfect matching.*

Proof. Suppose G has a perfect matching. We show that player 2
has a winning strategy.

To win the game player 2 fixes a perfect matching M and carries
that in his head. Player 1 chooses a vertex x to start the game.
There is a unique edge $e \in M$ that contains x. Player 2 chooses
the other endpoint of e to extend the path.

A perfect matching in a
graph is a matching with $n/2$
edges; so it covers all the
vertices.

At any point in the game when player 1 enters a new edge of M player 2 chooses the other end of that edge. Since M is perfect this is a winning strategy for player 2.

The exercise below takes care of the converse.

This proves the theorem. $\square$

Exercise 2.70

Suppose that G has no perfect matching. Show that player 1 has a winning strategy.

2.18.2   Grundy values

When it is a player's turn he is faced with a <u>position</u> in the game. Assume that the positions in a game form a **DAG** when there is an arc from position x to position y if y is reached from x in <u>one</u> move (by either player). [56]

Give each position a <u>Grundy value</u> defined as follows. Each sink in the **DAG** has Grundy value 0. Let x be a position with some outgoing arcs. The Grundy value of x is the smallest element in $\mathbb{N} \cup \{0\}$ that is not a Grundy value of any of its out-neighbors.

Let s denote the starting - position — ie — s is the position in which player 1 has to make his first move. Call the Grundy value of s the Grundy value of the game.

PLAYER 1 HAS A WINNING STRATEGY IF AND ONLY IF THE GRUNDY VALUE OF THE GAME IS NOT 0.

To see that assume that the Grundy value of s is not 0. By the definition of the Grundy value there must be a position y that player 1 can reach which has Grundy value 0. Player 1 makes that move. Since the Grundy value of y is zero player 2 can only make moves to positions that have a non-zero Grundy value. — So — the game ends when player 1 makes the final move to a sink.

A natural question is whether player 1 has a winning strategy.

[56] We want to consider only <u>finite</u> games; if the digraph has a directed cycle a game can go on forever.

A sink in the **DAG** is a position that ends the game.

My teacher once said that we could try to simulate chess with a **DAG**; but it's made hard by rules that involve the <u>history</u> of the game (like 'castling'). "And then — you need to deal with a case where you offer a draw and your opponent gets a red face and resigns!"

> Input: A graph G and a game (ie a rule which defines legal moves).
>
> Output: Decide whether the Grundy value of a the game is 0.

For example, there is a polynomial-time algorithm to decide if the Grundy value of SNAKE is zero; just figure out if the graph has a perfect matching.

2.18.3   De Bruijn's game

De Bruijn shows that is can be quite difficult to find a winning move.

To decide a game is to decide whether its Grundy value is 0. If the Grundy value is > 0 then player 1 has a winning move. In this section we show a game in which it is not easy to find that winning move.

Consider this game played on the set $[n]$. During the game two players alternate and scratch out certain numbers of $[n]$. A number can be scratched out only once and when no number is left the player who has to make a move loses the game.

The rule to scratch out numbers is this. When it is a player's turn he chooses a number that has not been scratched out before. The move scratches out this number and all its divisors (ie those divisors that were not scratched out already).

So the number 1 gets scratched out at the first move.

PLAYER 1 HAS A WINNING STRATEGY.

To see that suppose player 1 chooses 1. This does not really change the game — that is — if now player 2 wins the game then player 1 could have made that winning move instead of playing 1! — In other words — player 1 wins the game.

In this game player 1 has a 'waiting move.'

Player 1 wins the game but finding the winning move is not feasible —say — when $n > 100$.

Exercise 2.71

The NIM - game is played with some piles of stones. When it is a player's turn he must choose one nonempty pile and remove some stones from it. When no pile has any stone left the player that has to make a move loses the game.

Let there be k piles and let the number of stones in these piles be

$$n_1 \quad \cdots \quad n_k.$$

The Grundy value of this game is the nim - sum

$$n_1 \quad \oplus \quad \cdots \quad \oplus \quad n_k.$$

To obtain the nim - sum add the n_i with the following addition rule

$$a \oplus b \;=\; \min\{k \in \mathbb{N} \cup \{0\} \mid \forall_{a' < a} \;\; \forall_{b' < b}$$
$$k \neq a' \oplus b \;\; \text{and} \;\; k \neq a \oplus b'\}.$$

Another way to obtain $a \oplus b$ is to write a and b in binary and then to add them up bit - by - bit without using a carry.

Exercise 2.72

In this game the board is a forest F of k rooted trees. On the root of each tree lies a coin. A legal move chooses one tree in the forest and moves the coin in this tree to a point that is further away from the root.

Let d_i denote the maximal distance of any point in $T_i \in F$ from the root. The Grundy number of this game is

$$d_1 \quad \oplus \quad \cdots \quad \oplus \quad d_k.$$

2.18.4 Poset games

POSET GAMES ARE PLAYED ON A POSET. Two players play a game on a poset $(P, \leqslant)$. When it is his turn a player selects an element x of P; this removes x and all elements $y \geqslant x$.

Example 2.116. 1. Clearly NIM is a poset game: the poset is a union of chains (the piles).

2. HACKENDOT is a game played on a forest of rooted tree. The selection as a move of a vertex removes all vertices that are on the path from the vertex to a root. (When the forest is a tree then player 1 wins.)

J. Úlelha, A complete analysis of Von Neumann's Hackendot, *International Journal of Game Theory* **9** (1980), pp. 107–113.

Deuber and Tomassé show that the Grundy value of an N - free poset - game can be computed in $O(n^4)$ time.

W. Deuber and S. Thomassé, Grundy sets of partial orders. Technical report, University Bielefeld, 1980.

2.18.5   Coin - turning games

LET A BOARD BE $[n]$ WITH A COIN ON EVERY ELEMENT THAT IS SHOWING head OR tail.   A move is a turn of two coins subject to the condition that the right - most coin of the two turns from head to tail.

Write the numbers $1 \cdots n$ from left to right.

As usual, when a player can't make a legal move he loses the game.

Exercise 2.73

Suppose there is only one coin that is showing head. Show that the game is equivalent to NIM with one pile of stones. How many stones are there in the pile of the NIM-game?

HINT: If there is only one coin that shows head then a legal move is the same as 'shifting the head to the left.' So, it is equivalent to NIM with one pile of stones. The number of stones on the pile in NIM is the number of positions the head can move to the left; if it is in position k then it can move $k - 1$ positions to the left. (If the head is in position 1 player 1 loses the game; accordingly $g(1) = 0$.)

LET'S TAKE A BOLD STEP and see what happens if we simulate the coin - turning game by a NIM - game that has one pile for every head in the set.

Let $A \subseteq [n]$ be the set of elements where the coin is showing head. Simulate this game by a NIM - game; with a pile of stones for every $a \in A$. The number of stones in a pile — say $g(a)$ — is the number of positions it can move to the left.

CLAIM: THE TWO GAMES HAVE THE SAME GRUNDY VALUE:

$$\bigoplus_{a \in A} g(a).$$

Proof. A difficulty arises only when two coins are turned that are both head. In a 'perfect' simulation of the coin - turning game the two piles would 'disappear.' The two piles in the NIM - game become of equal size   that is — they have the same number of stones and so the sum of their Grundy values equals 0.

This proves the claim.   □

Exercise 2.74

Suppose the coin - turning game is played on an $n \times m$ - grid. Call (n, m) - corner of the grid the North - East. One each point of the grid lies a coin which shows head or tail.

A legal move turns all coins of a subgrid subject to the condition that the North - East - corner of the subgrid turns from head to tail.

Show that the Grundy value can be expressed as

$$\bigoplus_{a \in A} g(a), \qquad (2.44)$$

where A is the set of vertices of the grid on which the coin shows head. (What is $g(a)$?)

Exercise 2.75

Let $(P, \preceq)$ be a poset. Define the coin - turning game analogously to the above. Show that there is an efficient way to decide if player 1 wins this game.

HINT: Introduce a turning set for each element in the poset. All coins in a turning set $T(a)$ turn when a player turns the coin in a. In this game we assume that each element a is the unique maximal element of its turning set $T(a)$.

A legal move flips all coins of a turning set $T(a)$ provided that the coin in a shows head. Show that the Grundy value of the game can be expressed as in (2.44).

2.18.6 Nim - *multiplication*

H. W. Lenstra, Jr., NIM - multiplication. Technical Report Institute des Hautes Etudes Scientifiques IHES/M/78/211. Research supported by the Netherlands Organization for the Advancement of Pure Research (Z.W.O), 1978.

This paper defines a game as follows.

This definition should end with an exclamation mark!

Definition 2.117. A game is a set.

To explain this definition a game is identified with its initial position and a position is a set of options. Each option is again a position. — So — an element of a game is a position (a set of options) and every element of a position is again a position. In this section all elements of a set are sets.

To play a set S Player 1 chooses an element of S — say — S'. Then Player two chooses an element S'' of S' and so on. The player who needs to choose an element from an empty set loses the game. We assume that this will — eventually — occur, after a finite number of moves.

While we're at it let's define the Grundy number of a game A as $g(\varnothing) = 0$ and

$$g(A) \quad = \quad \min\{\, k \mid k \neq g(\ell) \quad \text{for} \quad \ell \in A \,\}$$

Then player 1 has a winning strategy if and only if the Grundy value of the game is not zero.

The sum of two game A and B is the game

$$A + B \quad = \quad \{\, a + B,\, b + A \mid a \in A \quad \text{and} \quad b \in B \,\}$$

We assume that you are familiar with the sum - theorem.

Theorem 2.118 (The sum theorem). *The Grundy value of the sum of two games* $A + B$ *is*

$$g(A + B) \quad = \quad g(A) \quad \oplus \quad g(B),$$

where $\oplus$ *is the* **NIM** *- addition:*

$$\alpha \oplus \beta \quad = \quad \min\{\,k\,|\,k \neq \alpha' \oplus \beta \quad and \quad k \neq \alpha \oplus \beta'$$
$$for\ all\ \alpha' < \alpha\ and\ \beta' < \beta.\,\}.$$

LET US DEFINE THE PRODUCT of two games as

$$A \times B \quad = \quad \{\,(a \times B) + (A \times b) + (a \times b)\,|\,a \in A \quad and \quad b \in B\,\}.$$

Then the Grundy value of the product is

$$g(A \times B) \quad = \quad g(A) \circ g(B)$$

where the product $n \circ m$ of two numbers is the smallest number

$$different\ from \quad (n' \circ m) \oplus (n \circ m') \oplus (n' \circ m')$$
$$for\ all\ n' < n\ and\ m' < m. \quad (2.45)$$

We want no zero divisors:

$$(n - n') \circ (m - m') \neq 0.$$

So the Grundy value of $n \circ m$ is the smallest number different from all

$$(n' \circ m) \oplus (n \circ m') \oplus$$
$$(n' \circ m').$$

SUPPOSE WE PLAY THE PRODUCT GAME with two naturural numbers n and m . After t moves the position looks like

$$(a_1 \circ b_1) + (a_2 \circ b_2) + \cdots + (a_{2t+1} \circ b_{2t+1}).$$

For any pair a, b the term $(a \circ b)$ may appear many times but only the parity of the number of occurrences of $(a \circ b)$ is of interest.

A <u>legal move</u> is to replace a pair — say (a, b) — with three pairs

$$(a' \circ b) + (a \circ b') + (a' \circ b'),$$

where $a' < a$ and $b' < b$.

$(\mathbb{N}, \oplus, \circ)$ is a field of characteristic <u>two</u>.

If one of the three new term is already in the product then the two equal terms cancel each other out because the sum of two equal games is zero.

We can represent the positions of the $n \circ m$ - game by a rectangular $n \times m$ - grid. In each point of the grid there is a coin showing red or blue (head or tail).

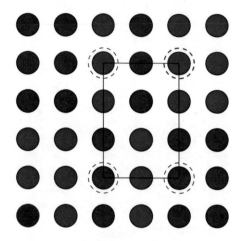

The rule for a legal move is as follows. Choose a rectangle of which the North - East corner is red. Switch the color at the 4 corners of the rectangle.

Theorem 2.119 (The product theorem). *The Grundy number of a product game is the sum over all the red nodes of the nim - product of the two coordinates.*

IN HIS PAPER (Exercise 4) Lenstra describes an algorithm to calculate the NIM - product of two numbers $n \circ m$.

2.18.7 P_3 - Games

Definition 2.120. Let G be a graph. A set $S \subseteq V$ is P_3 - convex if

$$\forall_{x \notin S} \quad |N(x) \cap S| \quad < \quad 2.$$

Notice that this defines an <u>alignment</u> — that is

1. $\varnothing$ and V are P_3-convex

2. if A and B are P_3-convex then $A \cap B$ is P_3-convex.

Exercise 2.76

Let $\mathcal{L}$ be the collection of P_3-convex sets. Define the <u>hull</u> <u>operator</u> $\sigma : 2^V \to \mathcal{L}$ by

$$\sigma(A) \quad = \quad \text{the smallest } P_3\text{-convex set that contains } A.$$

Show that this is a proper definition.

Two players play the <u>P_3-game</u>. The board is a graph. In a move certain vertices of the graph get labeled. Initially the set of labeled vertices $S = \varnothing$.

A player selects an unlabeled vertex x. This changes the set of labeled vertices as follows.

$$S \quad \leftarrow \quad \sigma(S + x)$$

(The move labels all vertices of $\sigma(S + x)$.)

(At each point of the game prior to a move the set of labeled vertices S is P_3-convex.)

Exercise 2.77

Prove the following theorem.

Theorem 2.121. *There exists an* $O(n^2)$ *algorithm to decide the* P_3 *- game on paths.*

HINT: Compute the Grundy value of the game played on every subpath — using dynamic programming.

In the <u>connected P_3 - game</u> the moves are restricted so that the set of labeled vertices S must induce a connected subgraph.

Wing Kai Hon, Ton Kloks, Fu-Hong Liu, Hsiang-Hsuan Liu and Tao-Ming Wang, P_3 - *games*. Manuscript on arXiv: 1608.05169, 2016.

Exercise 2.78

1. Player 1 wins the connected P_3 - game on P_n if and only if

$$n \neq 2.$$

2. Player 1 wins the connected P_3 - game on the cycle C_n if and only if

$$n = 2 \mod 3.$$

Exercise 2.79

1. Show that there is a polynomial - time algorithm to decide the connected P_3 - game on trees.

2. Show that there is a polynomial - time algorithm to decide the P_3 - game on cographs.

3. The <u>ladder</u> is the Cartesian product $P_2 \times P_n$. Show that Player 1 wins the connected P_3 - game on the ladder if and only if

$$n = 0 \mod 6.$$

2.18.8   Chomp

It is also called 'the take-away game.'

TWO PLAYERS PLAY A GAME ON A GRAPH. The name of the game is CHOMP. When it is his turn a player removes a vertex or edge of the graph. [57] The game ends when there are no more vertices or edges left. The game ends when there is no graph left to play with.

[57] The removal of a vertex also removes all edges that are incident with it.

Exercise 2.80

Show that player 1 loses the game on a triangle. Show that the Grundy value for Chomp on K_n equals $n \mod 3$.

Exercise 2.81

Show that there is an efficient way to decide the winner of a game of chomp on K_n^+ which consists of a clique with n vertices and one extra vertex that is adjacent to exactly one vertex in the clique.

A useful tool to compute the Grundy value of this game is the flipping lemma. A flip is an automorphism $\sigma : V(G) \to V(G)$ which satisfies

1. for every $x \in V$ $\{x, \sigma(x)\} \notin E$

2. $\sigma^2 = \sigma$.

The kernel of a flip is the set

$$\{x \mid \sigma(x) = x\}.$$

Lemma 2.122 (The flipping lemma). *The Grundy value for chomp on G equals the Grundy value on any kernel of a flip.*

Kandhawit and Ye — extending older results of Draisma and Van Rijnswou for the Grundy value of forests — showed that for bipartite graphs the Grundy value equals

$$\phi(G) = n_2 + 2 \cdot m_2, \tag{2.46}$$

where n_2 and m_2 are the numbers of vertices and edges of the graph modulo two.

Exercise 2.82

1. Show that an even wheel has a flip with kernel P_3. — Consequently — even wheels have Grundy value 1.

2. Show that also odd wheels have Grundy value 1.

Exercise 2.83

Is there a polynomial time - algorithm to decide the winner of a game of chomp played on a cograph?

HINT: Every cograph is either one vertex or the join or the union of two smaller cographs. — Clearly — the Grundy value of the graph is the nim - sum of the Grundy values of its components.

Problem Formulations

TO COMMUNICATE problems in mathematics one makes use of logic. In this section we discuss two languages that can be used to express 'most' problems in algorithmic graph theory.

3.1   Graph Algebras

Many sets of graphs can be described via a 'gluing' or a 'bridging' procedure.

Given two graphs — say G_1 and G_2 — in a bridging operation one creates a new graph by adding edges between certain subsets of vertices of G_1 and G_2. Examples of classes of graphs that fall into this category are trees, cographs, and distance-hereditary graphs. [1]

In a gluing operation one builds a larger graph from G_1 and G_2 by identifying certain vertices of G_1 and G_2. An example of a class of graphs that falls in this category is the class of chordal graphs.

The languages that we will discuss next allow us to formulate both the membership of graphs in these classes and most problems that are in NP.

[1] A graph is distance hereditary if for any two vertices in the graphs, all chordless paths that connect them, have the same length.

3.2   Monadic Second – Order Logic

Monadic second–order logic is a logical language — that can be used to express properties of graphs.

— For example — according to Definition 1.5 — we can express that a graph G is connected as follows.

$$\forall_{A \subseteq V}\ \forall_{B \subseteq V}$$
$$(A \neq \varnothing \wedge B \neq \varnothing \wedge A \cap B = \varnothing \wedge V = A \cup B)\ \Rightarrow$$
$$\exists_{a \in V} \exists_{b \in V}\ (a \in A \wedge b \in B \wedge \{ a, b \} \in E).\quad (3.1)$$

— As you can see — in this language we use <u>quantifiers</u> — that is — the symbols

$$\forall \quad \text{and} \quad \exists.$$

The quantities — that we quantify over — are subsets of V and elements of V. [2] The rest of Formula 3.1 is called the <u>expression</u> of the sentence.

[2] That's why the language is called 'monadic.'

3.2.1   Sentences and Expressions

A sentence consists of

1. Logical symbols:

   (i) Parentheses (and)

   (ii) Connective symbols $\Rightarrow$, $\wedge$, $\vee$, and $\neg$

   (iii) Variables, say x, y, A, any natural number of them

2. and Parameters:

   (a) Quantifier symbols $\forall$ and $\exists$

   (b) Equality symbol $=$

   (c) Constants symbols, eg, $\varnothing$, and any natural number

(d) Predicate symbols — that is — any number of functions

$$f : V^k \rightarrow \{ \, \text{true} \, , \, \text{false} \, \} \quad \text{for } k \in \mathbb{N}$$

— for example — $(\{a, b\} \in E(G))$. [3]

For simplicity we assume that sentences are well–formed — that is — the parentheses are placed so that they break down the formulas into their atomic parts. These parts have no connective nor quantifier symbols.

3.2.2   Quantification over Subsets of Edges

A version of Monadic Second–Order Logic — where quantification over subsets of edges is allowed — turns out to be important for algorithms on graphs of bounded treewidth (see Section 4.2).

Definition 3.1. A property is expressible in MS_1 if it can be expressed in a monadic second–order formula with quantification over vertices and subsets of vertices.
A property is expressible in MS_2 if it can be expressed in a monadic second–order formula with quantification over vertices, edges and subsets of vertices and edges.

Exercise 3.1

Show that having a <u>Hamiltonian Cycle</u> is a problem that can be expressed in MS_2. Can you also express it in MS_1 ?
Hint: A Hamiltonian cycle is a of edges that form a cycle — and — that covers all vertices.

Remark 3.2. Courcelle proved the following two theorems — which we mention here for completeness' sake.

Theorem 3.3. All MS_1 – sentences can be evaluated in $O(n^3)$ time on graphs bounded rankwidth. [4]

[3] The number of arguments of a predicate is called its arity.

[4] The class of graphs of bounded rankwidth is a parametrization of the class of distance-hereditary graphs.

These algorithms are fixed-parameter algorithms, parametrized by the rankwidth of the graphs.

Historically the next theorem came first.

Theorem 3.4. All MS_2*-sentences can be evaluated in linear time on graphs of bounded treewidth.*

These are fixed-parameter algorithms, where the parameter is the treewidth of the graph.

Exercise 3.2

The language MS_2 is more powerful than MS_1. Therefore, if we let MS_i stand for the class of graphs, for which all MS_i-problems can be solved in universal-constant polynomial time, then we have

$$MS_2 \subseteq MS_1.$$

— Find other problems — that are expressible in MS_2 but not in MS_1.

Recent Trends

As this is the final chapter of this overview
we should have a look at the 'recent' developments in graph
algorithmics. [1] To keep this brief short let me select one
topic — namely <u>treewidth</u> — and use that as a chassis to
explain various 'recent' concepts.

4.1   Triangulations

A triangulation of a graph is an embedding of it in a chordal
graph.

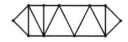

Figure 4.1: A chordal graph

4.1.1   Chordal Graphs

Definition 4.1. A graph is <u>chordal</u> if it has no chordless
cycle [2] of length more than 3.

We could say that chordal graphs are $\{C_n\}$–free graphs,
(for all $n \geqslant 4$), except that this notation looks a bit weird
(since it is not finite). Trees are well–known examples of
chordal graphs and in general — one could say — that chordal
graphs have a 'tree‑like'–structure.

The way in which the structure of chordal graphs resembles
that of trees is best conveyed via their minimal separators.

Exercise 4.1

Prove that every connected
chordal graph, with at least
two vertices, has a simplicial vertex, that is, a vertex whose neighborhood is a
clique.
Hint: Consider a feasible partition, see Definition 2.69 on
page 76.

Lemma 4.2. *A connected graph is chordal if and only if all its minimal separators are cliques.* [3]

Proof. First, assume that G has a minimal separator S with two nonadjacent vertices — say a and b. Let S be a minimal $x|y$- -separator — and let C_x and C_y be the components of $G - S$ that contain x and y. Then every vertex of S has a neighbor in C_x and in C_y.

Two chordless $a \sim b$ – paths — one with its internal vertices in C_x and the other with its internal vertices in C_y — form a chordless cycle in G of length at least 4. — So — when G has a minimal separator that is not a clique then G is not chordal.

Now assume that every minimal separator in G is a clique. When G is a clique itself we are done, since cliques are chordal graphs. We proceed by induction on the number of vertices in the graph. [4]

Let S be a minimal separator and let $C_1, \cdots, C_t$ be the components of $G - S$. We claim that every minimal separator in each graph $C_i \cup S$ is a clique.

— To see that — let S' be a minimal $x|y$-separator in $C_1 \cup S$. Since S is a clique it can be contained in at most one component of

$$(C_1 \cup S) \setminus S'.$$

It follows that S' is a minimal separator in G — and so — S' is a clique.

By induction — on the number of vertices — each graph $C_i \cup S$ is chordal — and since any chordless cycle of length at least 4 would be contained in one $C_i \cup S$ — such a chordless cycle cannot exist in G.

This proves the lemma.   □

[3] We exclude disconnected graphs, because, those have $\varnothing$ as a minimal separator, and $\varnothing$ is not a clique.

[4] Alternatively, we could proceed by induction on the number of minimal separators in the graph.

USUALLY TREES HAVE LEAVES. In the language of chordal graphs leaves are called simplicials.

Definition 4.3. A vertex is <u>simplicial</u> if its neighborhood is either $\varnothing$ or a clique.

Lemma 4.4. *Every chordal graph has a simplicial.*

Proof. If G is a clique we are done since then every vertex is a simplicial.

Consider a feasible partition $\{X, S, C\}$.[5] Then S is a clique since it is a minimal separator in a chordal graph. By induction on the number of vertices in the graph, $G[X]$ has a simplicial vertex. Since S is a clique this is also a simplicial in G.

This proves the lemma. $\square$

[5] See Section 2.9.1.

Since cycles have no simplicials we have the following characterization.

A graph is chordal if and only if every induced subgraph of G has a simplicial.

Equivalently — we have —

Corollary 4.5. *A graph is chordal if and only if it has a <u>perfect elimination order</u> — that is — an ordering of its vertices — say*

$$x_1 \quad \cdots \quad x_n \qquad\qquad —$$

such that

Abuse coming up ! !

$$\forall_i \quad x_i \ \text{is simplicial in} \ G[x_i \cdots x_n].$$

Exercise 4.2

Prove that every chordal graph that is not a clique has at least two simplicial vertices.

4.1.2 Clique – Trees

From a computational point of view the clique – tree of a chordal graph says it all.

Definition 4.6. A <u>clique tree</u> of a graph G is a pair (T, C) where T is a tree and C is the set of all maximal cliques in G. Furthermore, there is a bijection [6] $V(T) \to C$ which satisfies the property [7]

> for each vertex $x \in V(G)$ the maximal cliques that contain x form a <u>subtree</u> of T under the bijection.

[6] In future we'll simply <u>identify</u> each vertex of T with one maximal clique of G.

[7] Let's call this the <u>subtree property</u>.

Theorem 4.7. *A graph is chordal if and only if it has a clique – tree.*

Proof. Assume G has a clique – tree (T, C). Consider a clique $C \in C$ that is a leaf of T. We contradict the maximality of C when we assume that every vertex of C is also an element of the only neighbor of C in T. Therefore — by the subtree – property — there is a vertex in C that appears in no other element of C. — That vertex — is a simplicial of G.

Notice that having a clique – tree is a hereditary property — that is — if a graph G has a clique – tree then so does every induced subgraph of G. This shows that G is chordal, since every induced subgraph has a simplicial.

Assume G is chordal. When G is a clique it has a clique – tree and then we are done. Otherwise — let x be a simplicial of G. By induction on $|V(G)|$ the chordal graph $G - x$ has a clique - tree, say (T', C').

Since x is simplicial its neighborhood $N(x)$ is a clique. Let $P \in C'$ contain $N(x)$. Create a new node for $N[x]$ and attach it in T to P. It is readily checked that this creates a clique – tree for G.

This proves the theorem. $\qquad\qquad\square$

Exercise 4.3

Show that if a graph has a clique-tree then so does every induced subgraph.

Exercise 4.4

Construct a clique–tree for the graph in Figure 4.1 on Page 129. Which chordal graphs have a clique–tree that is a path?

HINT: Chordal graphs that have a clique–tree which is a path are called interval graphs. Consider the intersection graph of a set of intervals on the real line.

Exercise 4.5

Show that a graph is chordal if and only if it is the intersection graph of a set of subtrees in a tree. By that we mean that there exists a tree T and a collection of n subtrees of T

$$\{ T_x \mid x \in V(G) \}$$

such that

$$\{ x, y \} \in E(G) \quad \Leftrightarrow \quad V(T_x) \cap V(T_y) \neq \varnothing .$$

HINT: Assume G is chordal. For a vertex $x \in V(G)$ consider the subtree T_x of all the maximal cliques in the clique–tree T that contain the vertex x.

Exercise 4.6

Let G be a connected chordal graph and let $(T, \mathcal{C})$ be a clique-tree for G. Let

$$\mathcal{S} = \{ C_i \cap C_j \mid \{ C_i, C_j \} \in E(T) \}. \qquad (4.1)$$

Show that $\mathcal{S}$ is the set of minimal separators of G.

4.2   Treewidth

Definition 4.8. A <u>chordal embedding</u> or <u>triangulation</u> of a graph G is a chordal graph — say H [8] — with

$$V(H) = V(G) \quad \text{and} \quad E(G) \subseteq E(H). \tag{4.2}$$

Notice that every graph has a chordal embedding — just add all edges to G to make it a clique. The triangulation is <u>minimal</u> if the removal of an added edge creates a chordless cycle. [9]

The objective of the treewidth problem is to find a chordal embedding with smallest clique number.

Definition 4.9. The <u>treewidth</u> of a graph $G = (V, E)$ is defined as

treewidth(G) =

    $\min \{ \omega(H) - 1 \mid H \text{ is a chordal embedding of } G \}.$ (4.4)

We use tw(G) to denote the treewidth of G .

Treewidth of Claw – Free Graphs

Computing the treewidth of a graph is NP–complete . — However — it is solvable in polynomial time for many special classes of graphs.

As an example — since we are already a bit familiar with the structure of claw–free graphs — let's have a quick look at the computational complexity of treewidth for claw–free graphs.

Arnborg et al. showed that the treewidth problem remains NP–complete for bipartite graphs — and also for —

<u>cobipartite graphs</u> . [10]

Exercise 4.8

Show that the treewidth problem remains NP–complete when restricted to claw–free graphs .

[8] Hi!

[9] which has to be a 4-cycle

Figure 4.2: A chordal embedding of C_5 : The dotted lines are added in the embedding.

Exercise 4.7

Show that the number of minimal triangulations of the cycle C_n is the Catalan number Cat_{n-2} . That is, it satsifies the recurrence

$$Cat_{n-2} = \frac{4 \cdot n - 10}{n - 1} \cdot Cat_{n-3} \tag{4.3}$$

with $Cat_0 = 1$. So, the number of minimal triangluations of C_5 is 5.

[10] A graph is cobipartite if its complement is bipartite.

4.2.1   Treewidth and brambles

LET'S START WITH THE DEFINITION OF A BRAMBLE.

Definition 4.10. Let G be a graph. Two subsets $A, B \subseteq V$ touch
if $A \cap B \neq \emptyset$ or there exist $a \in A$ and $b \in B$ with $\{a, b\} \in E$.

Definition 4.11. Let G be a graph. A bramble $B = \{B_i\}$ is a
set of subsets $B_i \subseteq V$ whis satisfies the following properties.

1. each subset B_i induces a connected subgraph of G

2. Each pair B_i and B_j touch.

NOW LET'S DEFINE THE ORDER OF A BRAMBLE.

Definition 4.12. Let $B = \{B_i\}$ be a bramble. The order of B is
the minimal number of elements in a hitting set for B.

A set Z is a hitting set for $\{B_i\}$ if $Z \cap B_i \neq \emptyset$ for all $B_i \in B$.

AND NOW WE DEFINE THE BRAMBLE NUMBER OF A GRAPH.

Definition 4.13. The bramble number of a graph G is the maximal order of a bramble in G.

We denote the bramble number of G by $b(G)$.

Exercise: What is the bramble number of an independent set? What is the bramble number of a clique? How to compute the bramble number of a graph from the bramble numbers of its components? Design an algorithm to compute the bramble number in cographs.

We write $tw(G)$ for the treewidth of G.

The reason we did all that is the following theorem proved by
Seymour and Thomas in 1993.

Theorem 4.14 (Seymour and Thomas). *For any graph G the
following equality holds.*

$$tw(G) + 1 \quad = \quad b(G)$$

There is an elegant proof of Theorem 4.14 by Bellenbaum and
Diestel.

P. Bellenbaum and R. Diestel, Two short proofs concerning treede-compositions, *Combinatorics, Probability, and Computing* **11** (2002), pp. 541–547.

The following lemma gives a short proof of the theorem for chordal graphs. (Perhaps it helps the reader to get a feel for brambles and for chordal graph.)

Lemma 4.15. *For any chordal graph* G

$$b(G) \quad = \quad \omega(G).$$

Proof. The graph is chordal so there is a tree T and a collection of subtrees of T — say $\{T_x \mid x \in V(G)\}$ — with the property that for any two subtrees T_x and T_y

$$V(T_x) \quad \cap \quad V(T_y) \quad \neq \quad \varnothing \qquad \Leftrightarrow \qquad \{x, y\} \quad \in \quad E(G)$$

Let $\{B_i\}$ be a bramble. By definition each $G[B_i]$ is connected. Let

$$T_i \quad = \quad \bigcup_{x \in B_i} T_x$$

Then T_i is a subtree of T.

Every pair B_i and B_j touch. This implies

$$V(T_i) \quad \cap \quad V(T_j) \quad \neq \quad \varnothing.$$

Since every pair of subtrees T_i and T_j share a point of T there exists a point c in T that is in every tree T_i. [11]

[11] The Helly property.

Let $C = \{x \in V \mid c \in V(T_x)\}$. Then C is a clique in G — and so — $|C| \leqslant \omega$. — Furthermore — the set C is a hitting set for the bramble.

We conclude that $b(G) \leqslant \omega(G)$ since any bramble has order at most $\omega(G)$.

To see that $\omega \leqslant b$ let M be a clique in G with ω vertices. Define a bramble $\{\{x\} \mid x \in M\}$. — Clearly — the order of this bramble is ω. So $\omega \leqslant b(G)$.

This proves the lemma. $\square$

Exercise 4.9

Show that any graph G satisfies

$$b(G) \leqslant tw(G) + 1.$$

Exercise 4.10

(a) Let G be a graph and let H be a minimal triangulation of G. Show that any bramble in G is a bramble in H of the same order.

(b) Show that for any graph

$$b(G) = tw(G) + 1.$$

Further reading

The paper below introduces brambles — although in this paper brambles are called 'screens.'

P. Seymour and R. Thomas, Graph searching and a minimax theorem for treewidth, *Journal of Combinatorial Theory, Series B* **58** (1993), pp. 239–257.

The paper that christens the concept as brambles is this.

B. Reed, Treewidth and tangles, a new measure of connectivity and some applications. In: Vol. 241 of *LMS Lecture Note Series*, Cambridge University Press, (1997), pp. 87–162.

4.2.2 Tree - decompositions

Graphs of treewidth k are exactly the graphs that have tree - decompositions of width k.

Definition 4.16. Let G be a graph. A tree - decomposition for G is a pair $(T, \{X_i\})$ where

1. T is a tree with a root

2. each node $i \in V(T)$ corresponds with a bag $X_i \subseteq V(G)$

3. every vertex of G is in a bag

4. every edge of G is contained in a bag

5. for each vertex $x \in V(G)$ the nodes i with $x \in X_i$ form a subtree of T.

The width of a tree - decomposition is the maximal size of a bag minus one.

A graph has treewidth k if and only if it has a tree - decomposition with width $\leqslant k$.

In 1996 H. Bodlaender designed a linear time algorithm to compute a tree - decomposition of minimal width for graphs of bounded treewidth.

Make a clique of every bag; this embeds the graph in a chordal graph with the tree - decomposition as a clique - tree.

H. Bodlaender, A linear time algorithm for finding tree - decompositions of small treewidth, *SIAM J. Comput.* **25** (1996), pp. 1305–1317.

Exercise 4.11

Show that a graph G has a <u>nice</u> tree - decomposition — that is — a tree - decomposition $(T, \{X_i\})$ which satisfies the following.

1. the width of T is $\operatorname{tw}(G)$

2. every node of T has at most two children

3. if a node i has one child j then $|X_i| = |X_j| + 1$ and $X_j \subset X_i$ or $|X_i| = |X_j| - 1$ and $X_i \subset X_j$

4. if a node i has two children p and q then $X_i = X_p = X_q$

5. T has at most $4n$ nodes (where $n = |V(G)|$).

4.2.3 Example: Steiner tree

AS AN EXAMPLE let us take a look at an algorithm that solves the Steiner tree problem on graphs of bounded treewidth. To be precise we take a close look at the following paper.

M. Chimani, P. Mutzl and B. Zey, Improved Steiner tree algorithms for bounded treewidth, *Journal of Discrete Algorithms* **16** (2012), pp. 67–78.

Partitions of a set

AS USUAL we start with something else.

Let S be a set. Say that S has k elements. We wish to enumerate all the partitions of S.

One way to do this is recursive and goes as follows. Choose an arbitrary element, say $a \in S$. Choose a part with j elements in $S \setminus a$ and put a in that part. Partition the remaining vertices in all possible ways.

A (mock) closed-form for B_k is

$$B_k = (1 + B)^{k-1}.$$

This algorithm gives the following formula for the number of partitions of a set with k elements.

$$B_k \quad = \quad \sum_{j=0}^{k-1} \binom{k-1}{j} \cdot B_{k-1-j}.$$

To make it all work out nicely we choose $B_0 = 1$.

These are the Bell numbers

$$1 \quad 1 \quad 2 \quad 5 \quad 15 \quad 52$$
$$203 \quad 877 \quad 4140 \quad \cdots$$

There is a simple algorithm to make a table of these numbers similar to Pascal's triangle. De Bruijn (in his book "Asymptotic Methods in Analysis") gives a nice asymptotic expression for $\ln(B_n)/n$ with a marvelous Big-Oh - term $O(\ln\ln n/(\ln n)^2)$. A more recent bound for the Bell numbers is

$$B_n < (0.792 n/\ln(n+1))^n.$$

Exercise 4.12

Design an algorithm that runs in $O(k \cdot B_k)$ time to make a list of all the partitions of a set with k elements.

Assume that we wish to enumerate all partitions of S where possibly one part is special.

This is done by adding a 'ghost - element' g to the set. Now enumerate all partitions of $S \cup \{g\}$. The part that contains the ghost element is the special part.

The algorithm gives the following formula for the number of partitions with a special part.

$$B_k^* \;=\; B_{k+1}.$$

The numbers $B_k^* - B_k$ appear in the second diagonal of the Bell triangle. (These are the number of partitions that have a special part.)

Steiner trees

Let G be a graph and let $\Omega \subseteq V$ be a subset of the vertices. The elements of Ω are called <u>terminals</u>. The <u>Steiner tree problem</u> is to connect the terminals by a tree in G with a smallest number of edges.

Equivalently: the Steiner tree problem asks for a connected subgraph of G with a smallest number of edges which contains all terminals. Clearly, this can only exist if the terminals are in a component of the graph.

Processing the tree - decomposition

To solve the Steiner tree problem we have a nice tree - decomposition $(T, \{X_i\})$ at our disposal. Let $i \in V(T)$. Let T_i be the subtree of T which is rooted at node i.

Chimani et al consider the problem where the edges have weights. We don't do that.

- $V_i \subseteq V(G)$ is the set of vertices that appear in bags of T_i

- $\Omega_i \subseteq \Omega$ is the set of terminals that appear in bags of T_i.

Let S be a Steiner tree in G. This induces a forest in $G[V_i]$. The only vertices of V_i that have neighbors in $V \setminus V_i$ are vertices of X_i. Assume that there is at least one terminal in $V \setminus V_i$. Then the terminals of $\Omega_i \setminus X_i$ are connected (in a forest) to some vertices in X_i. The algorithm stores at the node i the sub - forest of S on the vertices of X_i.

A forest on X_i is represented as a partition of X_i. One part of this partition may be special; a set of vertices that is not in the forest. The other parts represent the components of the forest.

A partition of X_i has a <u>cost</u>. The cost of a partition is the smallest number of edges in a Steiner forest in V_i which induces the partition.

The cost of a partition is the sum of the costs of the parts.

We may assume that every node in T is of one of the following types.

See Exercise 4.11.

1. a <u>start node</u> satisfies $|X_i| = 1$

2. an <u>introduce node</u> i has exactly one child j and the bags satisfy $X_j \subset X_i$ and X_i has exactly one vertex that is not in X_j

3. a <u>forget node</u> i has exactly one child j and the bags satisfy $X_i \subset X_j$ and X_j has exactly one vertex that is not in X_i

4. a <u>join node</u> i has two children p and q and the bags at these nodes satisfy $X_i = X_p = X_q$.

A node in the tree T is processed at a time after the completion of processing its children. Below we describe — for each type of node — the computational process.

Bottoms up!

Process at a start node

Let i be a start node — that is — X_i contains one vertex $x \in V(G)$. A table at i has the partitions of X_i with a possible special part. So at i we have a table with two entries $\{x\}$; in one of the entries $\{x\}$ is marked 'special.'

The special part contains the vertices of X_i that are not in the Steiner tree.

We need to supply a cost to each part of a table entry. When the vertex x is a terminal and a part that contains x is marked as a 'special' then the partition has cost ∞. In all other cases the partition has cost 0.

Process at an introduce node

Let i be an introduce node with a child j. Let x be the vertex that is in $X_i \setminus X_j$. Then

$$N[x] \quad \cap \quad V_i \quad \subseteq \quad X_i.$$

THE FIRST STEP of the process is to generate a table with all partitions of X_i with one part possibly marked as special. That is equivalent to the generation of all partitions of X_j with t̰w̰o̰ ghost elements — of which one is x.

The algorithm needs to compute a cost of each part of a partition P at the node i. This is obtained from the costs of partitions Q at node j that are <u>compatible</u> with P — as follows.

> some parts of Q connect with x into one part of P.

A partition Q of X_j is compatible with P if each part of Q either

- is equal to a part in P or

- is a subset of the part in P that contains x.

When a part of P is equal to a part of Q then its cost is copied. Let P_x be the part of the partition that contains x. Let Q be compatible with P and let

$$P_x \setminus x \quad = \quad \bigcup Q_i,$$

where $\{Q_i\}$ is the partition of $P_x \setminus x$ into parts of Q.

The cost of the part P_x is ∞ when $x \in \Omega$ and P_x is marked as special. — Otherwise — it is the smallest value — over all partitions Q that are compatible with P — of $\sum \mathrm{cost}(Q_i)$ **plus**

- ∞ when x is not connected to some Q_i

- the number of parts Q_i when they all connect to x.

> The only neighbors of x in V_i are vertices in X_i.

Exercise 4.13

Show that a table for the node i can be computed in $O(B_{k+2}^2 \cdot k)$ time where $k = \mathrm{tw}(G)$.

Exercise 4.14

Let i be an introduce node with child j and let $x \in X_i \setminus X_j$. Consider partitions of X_j with a 2-coloring on its parts. Let the parts of one color union with x to make up one part in a partition of X_i. Show that the number of 2-colored partitions (of a set with k elements) satisfies $B_0^{(2)} = 1$, $B_1^{(2)} = 2$ and

$$B_k^{(2)} \quad = \quad \sum_{j=0}^{k} \binom{k}{j} \cdot B_j \cdot B_{k-j}.$$

Show that $B_k^{(2)} \leqslant B_k \cdot B_{k+1}$.

Process at a forget node

NOT MUCH HAPPENS IN THIS CASE. We have $X_i = X_j \setminus v$ for some $v \in X_j$. Call two partitions of X_j equivalent if they are the same when the vertex x is removed. Compute the cost of an equivalence class (which is a partition of X_i) as the smallest cost of the elements in the class.

We leave it as an exercise to show that forget nodes are processed within $O(B_{k+2}^2 \cdot k)$ time where $k = tw(G)$.

Process at a join node

Let i be a join - node with two children p and q. We have for the sets of vertices

$$V_p \cup V_q \quad = \quad V_i \quad and \quad V_p \cap V_q \quad = \quad X_i$$

and for the sets of edges in $G[V_p]$ and in $G[V_q]$:

$$E_p \cup E_q \quad = \quad E_i \quad and \quad E_p \cap E_q \quad = \quad E_i.$$

A Steiner tree has edges in E_p or in E_q or in both — that is — we can split it up in two forests; one with edges in E_p and the other with edges in E_q. The monochromatic parts are a partition of X_i (with some part that is not used).

The algorithm needs to check if the two forests at p and q add up as a tree (to make one part of a partition at node i).

Consider all $2^{k-1}k^{k-2}$ trees in K_k where $k = |X_i|$ with a 2-coloring on the edges. Let the partition at p have the components of the tree with edges in color 1 and let the partition at q have the components of the tree with edges in color 2. (Both partitions may have a special part.) The cost of the tree is the sum of the costs of the partitions of p and q.

This yields an algorithm to compute a cost for all partitions at join - node i. To compute a table at node i it uses $O((2k)^{k-1}B_{k+1})$ time. The paper by Chimani et al improves on this. [12]

> [12] We have $B_k < (k/\ln k)^k$ (when $k \to \infty$).

Consider partitions P and Q at nodes p and q. Construct a graph on parts $\{P_i\}$ and $\{Q_i\}$ that make up a part of a partition at node i. The following algorithm checks if the union of $\{P_i\}$ and $\{Q_i\}$ is a proper part at node i.

> Exercise: $k!/B_k \to \infty$ and $k!/B_k^2 \to 0$ (as $k \to \infty$).

Construct a graph whose vertices are the parts in $\{P_i\}$ and the parts in $\{Q_i\}$. Two parts are connected by an edge if they share a vertex in X_i. This graph has $O(tw(G))$ vertices and edges. A depth - first - search on this graph detects whether it is connected and if there is any cycle in linear time.

In their paper Chimani et al show that this gives an algorithm that computes a table for a join - node in $O(B_{tw+2}^2 \cdot tw)$ time.

Conclusion

The cases described above add up to prove the theorem of Chimani, Mutzel and Zey.

> We are not aware of any (nontrivial) lowerbounds. — Perhaps — there is room for improvement.

Theorem 4.17. *There is an algorithm that takes $O(k \cdot B_{k+1}^2 \cdot n)$ time to solve the Steiner tree problem for graphs with treewidth k.*

Must - reads on Steiner trees

S. Dreyfus and R. Wagner, The Steiner problem in graphs, *Networks* **1** (1972), pp. 195–207.

A. Aitken, A problem in combinations, *Mathematical Notes* **28** (1933), pp. 18–23.

A. Marcus and G. Tardos proved the Stanley - Wilf conjecture in 2004. [13]

[13] ACTUALLY they proved the Füredi - Hajnal conjecture about $\{0,1\}$ - matrices.

Theorem 4.18. *For every permutation π there is a constant C such that the number of permutations in S_n that avoid π as a pattern is at most C^n.*

(The constant $C(\pi)$ is an exponential function of π for almost all permutations.)

A. Marcus and G. Tador, Excluded permutation matrices and the Stanley - Wilf conjecture, *Journal of Combinatorial Theory, Series A* **107** (2004), pp. 153–160.

Any permutation has a cycle decomposition. (That shows that $k! > B_k$ when $k \geqslant 3$; B_k is the number of permutation in which each cycle is ordered. The average number of cycles is the harmonic number $H_n \approx \ln(n)$) The number of permutations drops dramatically when we forbid a pattern.

Exercise 4.15

How many pairs (α, β) of permutations in S_n are there that have cycles C_α and C_β such that the union of $\alpha \setminus C_\alpha$ and $\beta \setminus C_\beta$ is a tree of cycles — that is — a <u>cactus</u>?

A connected graph is a cactus if any two cycles share at most one vertex. SO every block is an edge or a cycle.

4.2.4 Treewidth of Circle Graphs

Consider a circle in the Euclidean plane. A chord of the circle is a line segment that connects two points of the circle.

Definition 4.19. A <u>circle graph</u> is an intersection graph of a set of chords of a circle in the Euclidean plane. — That is — the vertex set of the circle graph is the set of chords of the circle and two vertices are adjacent whenever their chords intersect.

Figure 4.3: A circle and two chords in it. The circle graph corresponding to this model has two, adjacent vertices.

— As an example — we show that there is a nice algorithm that computes the treewidth of circle graphs.

To compute the treewidth of a graph we need to find a triangulation of it that minimizes the clique number. In a chordal embedding of a graph all minimal separators are cliques. — It follows that they are non−crossing.

Crossing Separators

Definition 4.20. One minimal separator S_1 <u>crosses</u> another one S_2 if there exist two components of $G - S_2$ — each containing a vertex of S_1.

We say that two minimal separators are <u>parallel</u> if they are noncrossing — that is — if neither crosses the other.

Exercise 4.16

Show that the crossing relation, on the set of minimal separators of a graph, is symmetric.

Exercise 4.17

Show that there is a $1-1$ correspondence between the set of minimal triangulations of a graph and the maximal sets of pairwise parallel minimal separators. — For example — in a chordal graph <u>all</u> minimal separators are pairwise parallel — and so — the graph has only one chordal embedding — namely the graph itself.

Minimal Separators in Circle Graphs

Consider an intersection model of a circle graph. [14] — That is — let C be a circle in the Euclidean plane and let G be a set of chords of C. We may assume that no two chords share an endpoint.

[14] Circle graphs can be recognized in 'almost' linear time.

Definition 4.21. A <u>scanline</u> is a chord of C that shares no endpoint with any chord of G.

Lemma 4.22. *Assume the graph* G *is connected. Let* S *be the set of chords of a minimal separator in* G. *There is a scanline* t *such that the elements of* S *are exactly the chords that cross* t.

Proof. Remove the chords from the intersection model that correspond with vertices of S. Then each part that remains connected corresponds to a component of $G - S$.

Let S be a minimal $a|b$-separator and let C_a and C_b be the components of $G - S$ that contain a and b. Choose a scanline t that separates the component C_a from C_b.[15] The chords

[15] The chords are straight line segments, therefore, there is a straight scanline that separates the convex hull of the chords of a component.

that cross t are exactly the chords of S.

This proves the lemma. $\square$

Consider a polygon P with $2n$ corners — one between every two consecutive endpoints of G.

Definition 4.23. A plane triangulation of P is a maximal set of noncrossing chords in P. [16]

Definition 4.24. Let T be a plane triangulation of P. The weight of a triangle in T is the number of chords in G that cross some sides of the triangle. The weight of the triangulation T is the maximal weight over the triangles contained in T.

Notice that, if a chord of G cross some side of a triangle, then it crosses exactly two sides of the triangle.

Figure 4.4: A plane triangulation of a 6-sided polygon.

An algorithm to compute the treewidth of circle graphs

To compute the minimal weight of all the triangulations of the polygon P we use dynamic programming.

Let ℓ be the number of corners of P [17] and let them be numbered

$$s_0, s_1, \cdots, s_{\ell-1}.$$

Denote the number of chords in G — that cross the line (s_i, s_j) — by $c(i, j)$.

Let $P(i, t)$ denote the sub–polygon with corners

$$s_i, s_{i+1}, \cdots, s_{i+t-1}$$

where we take indices modulo ℓ. Let $w(i, t)$ denote the weight of a minimal triangulation of $P(i, t)$.

[16] Let's call chords of P, diagonals

[17] If $n = |V(G)|$ then the circle graph model has n chords, with $2n$ endpoints. between any two consecutive endpoints, we have a corner of P. Thus, the polygon P has $\ell = 2n$ corners.

Organize the computation by increasing length of the sub-polygons. For $t = 2$ let $w(i, 2) = 0$ for all $i \in \{0, \dots, \ell-1\}$. Then we have — for $t \geqslant 3$ —

$$w(i, t) = \min_{2 \leqslant j < t} \max \{w(i, j), w(i+j-1, t-j+1), F(i, j)\}$$

$$(4.5)$$

where

$$F(i, j) = \frac{1}{2} \cdot (c(i, i+j-1) + c(i+j-1, i+t-1) +$$

$$c(i, i+t-1)). \quad (4.6)$$

That is so because every chord of G crosses — zero — or exactly two sides of every triangle.

Theorem 4.25. *The minimal weight of a triangulation of* P *can be computed in* $O(n^3)$ *time. The treewidth of* G *is*

$$\mathsf{tw}(G) = w(0, \ell) - 1.$$

Proof. There are $O(n^2)$ diagonals in P. A check whether a diagonal of P and a chord of G cross each other can be achieved in $O(1)$ time. Thus — to compute the numbers $c(i, j)$ of chords that cross a diagonal (s_i, s_j) takes $O(n^3)$ time.

By Equation (4.5) the minimal weight of a triangulation is obtained in $O(n^3)$ time.

To compute the treewidth of G we need to select a maximal set of parallel minimal separators. This problem is equivalent to finding a triangulation of P. The clique number of the minimal triangulation of G is the maximal number of chords that cross a triangle in the minimal triangulation of P.

This proves the theorem.   □

Exercise 4.18

Describe all minimal triangulations of the circle graph in the figure on Page 149. What is the treewidth of this graph?

I have not given you the definition of rankwidth [18] yet, but — for the record — I put down the following research problem here.

Exercise 4.19

Research problem:
Design a polynomial–time algorithm to compute the rankwidth of circle graphs.

David Chandler once showed me an algorithm, but, as far as I know, the details were not written down fully.

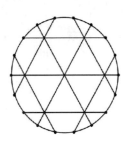 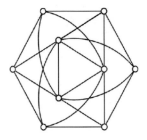

Figure 4.5: A circle graph and its model.

4.3 On the treewidth of planar graphs

BEFORE WE SAY ANYTHING ELSE let us mention that at present ie anno Domini 2021 the complexity of computing treewidth of planar graphs is OPEN.

In a remarkable paper Seymour and Thomas showed (in 1994) that treewidth of planar graphs can be approximated within a factor $3/2$. Their algorithm computes a decomposition in $O(n^4)$ time. In this chapter we describe their method.

LET G BE A GRAPH. In this chapter the definition of a graph is relaxed so that a graph may have multiple edges and loops. [19]

CONSIDER A TERNARY TREE T and a 1-1 map from the vertices of G to the leaves of T. [20] Identify an edge $e \in E(T)$ with a bipartition of $V(G)$ where two elements occupy a similar part if they appear in the same component of $T \setminus e$. [21]

The collection of (all) the classes of bipartitions of V over all edges of T defines a 'cross-free set-system' on $V(G)$. [22]

Definition 4.26. Let V be a finite set with at least two elements. Two subsets A and B (subsets of V) <u>cross</u> if

$$A \cap B \qquad A \setminus B \qquad B \setminus A \qquad V \setminus (A \cup B)$$

are all non-empty.

A family C of subsets of V is a <u>carving</u> if

C1 $\varnothing, V \notin C$

C2 no two elements of C cross

C3 C is maximal subject to the above.

---

[19] The dual of a plane graph may have loops and multiple edges. Without this relaxation the dual of a plane graph would not be a graph.

[20] A tree is <u>ternary</u> if every vertex has degree 1 or 3.

[21] The tree is called a 'routing tree' of the graph. Routing trees with small 'congestion' (ie carving width) are of importance for the design of telephone networks.

[22] A set-system C is <u>laminar</u> if for any two elements A and B of C at least one of the three sets $A \cap B$, $A \setminus B$ and $\overline{B \setminus A}$ is empty. Notice that 'being laminar' is a property that is independent of the ground set V. ('Being cross-free' is not.)

Exercise 4.20

Let C be a carving of a finite set V, $|V| \geqslant 2$. Show that there exists a ternary tree T and a 1-1 map $V(G) \to \text{leaves}(T)$ such that $C \in C$ if and only if there exists an edge e in T such that C is identified with the set of leaves in one of the two components of $T \setminus e$.

Let G be a graph. For $X \subseteq V$ let $\delta(X)$ denote the set of edges in G that have exactly one endpoint in X. [23]

Definition 4.27. Let G be a graph with at least two vertices. The width of a carving C of V is

$$\max \{ |\delta(X)| \mid X \in C \}.$$

The <u>carving width</u> of G is the smallest width of a carving of V.

[23] In this chapter we show that the carving width is computable in polynomial time for planar graphs.

Let $p : E \to \mathbb{Z}^{\geq 0}$. For $A \subseteq E$ let $p(A) = \sum_{e \in A} p(e)$. [24] The p-carving width of a graph with at least 2 vertices is the minimum over all carvings $\mathcal{C}$ of V of the maximum $p(\delta(A))$ for $A \in \mathcal{C}$.

We show below that the p-carving width of a connected planar graph is at least k if and only if either

1. there exists a vertex v satisfying $p(\delta(v)) \geq k$

2. it has an 'antipodality' of p-range at least k.

IN OTHER WORDS the minimal $k \in \mathbb{N} \cup \{0\}$ such that there is a carving $\mathcal{C}$ of $V(G)$ satisfying

$$\forall_{X \in \mathcal{C}} \quad p(\delta(X)) \quad \leq \quad k$$

equals the maximal $k \in \mathbb{N} \cup \{0\}$ such that G has an antipodality of p-range at least k or a vertex satisfying $p(\delta(x)) \geq k$. [25]

Next, we will show that there is a very easy algorithm (using the concept of 'round set') to find an antipodality of maximal p-range.

4.3.1 Antipodalities

A walk in a graph is a sequence

$$v_0 \quad e_1 \quad v_1 \quad e_2 \quad v_2 \quad e_3 \quad \cdots \quad e_k \quad v_k$$

with all $v_i \in V$ and all $e_i \in E$ and all $e_i = \{v_{i-1}, v_i\}$. The walk is closed if $v_0 = v_k$.

LET Σ BE A SPHERE [26] and let a graph G be embedded on Σ. Let $R(G)$ denote the regions of the embedding — that is — the maximal open sets (faces) in Σ that do not contain any vertex nor hit any edge.

Assume that G is not empty and let G^* denote the dual of G. Each vertex $v \in V$ is contained in exactly one region $r \in R(G^*)$; we denote $v^* = r$. Similarly, each region $r \in R(G)$ contains exactly one vertex $r^* \in V(G^*)$ (and that vertex is denoted r^*). For an edge $e \in E$ we let e^* be the unique edge of G^* that crosses e in Σ.

[24] $\mathbb{Z}^{\geq 0} = \mathbb{N} \cup \{0\}$. We could replace each edge of G by $p(e)$ parallel lines and compute the carving width of this auxiliary 'graph.' However it is our aim to find a 'strongly polynomial' algorithm for p-carving width — the timebound of our algorithm is a polynomial in $n + m$ (which gives a much better bound than a polynomial of the same degree in $n + \sum p(e)$).

[25] Slogan:

$$\text{min carvingwidth} = \text{max antipodality}$$

[26] A sphere is a balloon. The rubber edition was invented by Michael Faraday in 1824. An embedding of a graph is a drawing of it on a balloon such that edges only meet at endpoints. The regions of the drawing are the maximal connected areas of the balloon's surface that are not touched by the edges or vertices of the drawing. The regions are 'open sets:' if you take any point in a region, then a small enough disc with positive radius around the point will be contained in that region.

Let $p : E(G) \to \mathbb{Z}^{\geqslant 0}$. For a walk

$$v_0 \quad e_1 \quad \cdots \quad e_k \quad v_k$$

in the dual G^* define the p-length as

$$p(f_1) \quad + \quad \cdots \quad + \quad p(f_k)$$

where $f_i = e_i^*$.

Definition 4.28. An antipodality of p-range $\geqslant k$ is a function α with domain $E(G) \cup R(G)$ such that for all $e \in E$ $\alpha(e)$ is a subgraph and for all regions $r \in R$ $\alpha(r)$ is a nonempty subset of V satisfying

(A1) for each edge $e \in E(G)$ the subgraph $\alpha(e)$ does not contain an endpoint of e

(A2) if an edge e and a region r are incident then $\alpha(r) \subseteq V(\alpha(e))$ and every component of $\alpha(e)$ has a vertex in $\alpha(r)$

(A3) if $e \in E$ and $f \in E(\alpha(e))$ then every closed walk in G^* that contains e^* and f^* has p-length at least k.

INTERMEZZO: ROUND SETS

Let N be a graph with vertex set $V(N) = I$ and let M be a graph with a partition of its vertices

$$\{X_i \mid i \in I\}.$$

Assume

$$x \in X_i \quad \text{and} \quad y \in X_j \quad \text{and} \quad \{x, y\} \in E(M) \quad \Rightarrow$$
$$i \neq j \quad \text{and} \quad \{i, j\} \in E(N).$$

(There is a homomorphism $M \to N$.)

Definition 4.29. A set $Z \subseteq V(M)$ is round if

$$\forall_{\{i,j\} \in E(N)} \quad \forall_{x \in Z \cap X_i} \quad \exists_{y \in Z \cap X_j} \quad \{x, y\} \in E(M).$$

We show that there is a greedy algorithm to decide whether M has a nonempty round set.

Exercise 4.21

Prove the following. Let Z be round and let $Z \subseteq S$. Assume that there exists a vertex $x \in S$ that has no neighbors in $S \cap X_j$ for some $j \in N_N(i)$. Then $Z \subseteq S \setminus \{x\}$.

HINT: If $x \in Z \cap X_i$ then it must have a neighbor in $Z \cap X_j \subseteq S \cap X_j$ since Z is round.

Exercise 4.22

Define the bipartite graph H as follows. The two color classes of H are I and $V(M)$. A vertex $x \in V(M)$ is adjacent to $j \in I$ if $x \in X_i$ and i adjacent to j in N.

Design an algorithm that find a maximal round set in M. Your algorithm should run in

$$O(\quad |V(M)| \quad + \quad |E(M)| \quad + \quad |E(H)| \quad).$$

HINT: For $\{x, j\} \in E(H)$ let

$$d(x, j) \quad = \quad |N_M(x) \cap X_j|.$$

Construct a stack that contains the vertices $v \in V(M)$ that satisfy $d(v, j) = 0$ for some $j \in I$. Initialize $Z = V(M)$. At the end of the recursion described below Z will be a maximal round set of M.

During the recursion vertices of L are deleted from L and from Z. This needs an update of values $d(v, j)$ (since a vertex in M is deleted). To estimate the timebound describe in detail how the stack L and the values $d(v, j)$ are maintained — in other words — present the algorithm and prove its correctness. [27]

[27] A stack is a linear data structure in which elements are added or deleted only from the top-end.

Notice that the property of being round is maintained under unions — so — there is a unique maximal round set.

We show that there exists an $O(m^2)$ - algorithm to decide if a graph has an antipodality of p-range $\geq k$.

> Let G be a connected planar graph which is not empty. Let
> $p : E(G) \to \mathbb{Z}^{\geq 0}$ and let $k \in \mathbb{Z}^{\geq 0}$. For $e \in E(G)$ let $\phi(e)$ be
> the following subgraph. The vertices of $\phi(e)$ are $V \setminus e$ (that is;
> all vertices except the endpoints of e) and the edges of $\phi(e)$ are
> $f \in E$ for which $f \cap e = \emptyset$ and for which every closed walk in
> the dual G^* that contains e^* and f^* has p-length $\geq k$.

Lemma 4.30. *If there exists an antipodality of p-range $\geq k$
then there exists one — say α — such that $\alpha(e)$ is a union of
components of $\phi(e)$ for all $e \in E$.*

Proof. Let β be an antipodality of p-range $\geq k$. By definition $\beta(e)$
is a subgraph of $\phi(e)$.

For $e \in E$ define $\alpha(e)$ as the union of those components of $\phi(e)$
that intersect $\beta(e)$. For all regions r define $\alpha(r) = \beta(r)$.

The map α satisfies the first and third antipodality-condition (Definition 4.28). To check the second condition let $e \in E$ and $r \in R$ and
assume that e and r are incident. Then since β is an antipodality

$$\alpha(r) \quad = \quad \beta(r) \quad \subseteq \quad V(\beta(e)) \quad \subseteq \quad V(\alpha(e)).$$

Every component of $\alpha(e)$ contains a component of $\beta(e)$. This
implies that every component of $\alpha(e)$ intersects $\alpha(r)$.

This proves that α is an antipodality of p-range at least k. $\qquad\square$

We show that the set of components of $\phi(e)$ for all $e \in E$ can be
computed in $O(m^2)$ time.

Exercise 4.23

Let G be a connected plane graph with a dual G^*. Let $p : E \to \mathbb{Z}^{\geq 0}$.
For two vertices $x, y \in V(G^*)$ denote their p-distance in G^* as
$d^*(x, y)$. [28]

[28] The p-distance is the shortest p-length of a path (that runs between two vertices in the dual graph).

Let $e^* = \{x, y\}$ and $f^* = \{p, q\}$ be edges of G^*; $e^* \cap f^* = \emptyset$. There is a closed walk in G^* of p-length $< k$ if and only if either

$$d^*(x, p) + d^*(y, q) \quad < \quad k - p(e) - p(f) \quad \text{or}$$
$$d^*(x, q) + d^*(y, p) \quad < \quad k - p(e) - p(f).$$

Exercise 4.24

Design an algorithm to compute the following.

INPUT: A connected plane graph G with a dual G^*. A function $p : E \to \mathbb{Z}^{\geqslant 0}$ and integer $k \in \mathbb{Z}^{\geqslant 0}$.

Output: The graph $\phi(e)$ and a list of the components of $\phi(e)$ for every edge $e \in E(G)$.

Your algorithm should run in $O(m^2)$ time.

HINT: Frederickson shows that the all pairs shortest paths problem can be solved on planar graphs in $O(n^2)$ time.

We describe an algorithm to decide whether G has an antipodality of p-range $\geqslant k$. Below we prove the correctness.

The input is a connected plane graph G with a dual G^*, a function $p : E \to \mathbb{Z}^{\geqslant 0}$ and an integer $k \geqslant 0$.

1. For each edge $e \in E(G)$ compute C_e ie the set of components of $\phi(e)$.

2. Construct a graph M with vertices

$$V(M) \quad = \quad \bigcup_{i \in I} X_i,$$

where $I = E \cup R$ and a partition $\{X_i \mid i \in I\}$ of $V(M)$ is defined as follows

(i) For each $r \in R$ let

$$X_r \quad = \quad \{(r, x) \mid x \in V\}$$

(ii) For each $e \in E$ let

$$X_e \quad = \quad \{(e, C) \mid C \in C_e\}.$$

A vertex (e, C) is adjacent to a vertex (r, x) in M if [29]

$$e \subseteq \bar{r} \quad \text{and} \quad x \in V(C).$$

3. Use Exercise 4.22 to check if M has a nonempty round set. The graph N has vertex set $I = E \cup R$. The adjacencies in N running between edges and regions is defined by their incidence in G.

4. The graph G has an antipodality of p-range $\geqslant k$ precisely when M has a nonempty round set.

Remark 4.31. To compute a round set in M construct a bipartite graph H with color classes $V(M)$ and $V(N)$ as follows. When an edge e and a region r are incident then e is adjacent to every vertex in X_r and r is adjacent to every vertex in X_e.

For $v \in V(M)$ and $j \in V(N)$ such that $\{v, j\} \in E(H)$ define $d(v, j)$ as the number of vertices in X_j that is adjacent to v in M ie

$$d(v, j) = \begin{cases} |C| & \text{if } j \in R, \ v = (e, C), \ (e \in E, \ C \in C_e) \\ 1 & \text{if } j \in E, \ v = (r, x), \ (r \in R, \ x \notin j) \\ 0 & \text{if } j \in E, \ v = (r, x), \ (r \in R, \ x \in j). \end{cases}$$

Create a stack with elements $v \in M$ for which there is a $j \in N$ satisfying $d(v, j) = 0$. Repeatedly delete those elements from M and update the stack.

Theorem 4.32. *Let G be a connected planar graph and let $p : E \to \mathbb{Z}^{\geqslant 0}$ and $k \in \mathbb{N}$. There exists an $O(m^2)$ - algorithm that decides if G has an antipodality of p-range $\geqslant k$. Here $m = |V(G)| + |E(G)|$.*

Proof. To prove the correctness let $Z \subseteq V(M)$. Define, for $e \in E$ and for $r \in R$, a subgraph $\alpha(e)$ and a subset of vertices $\alpha(r)$:

$$\begin{aligned} \alpha(e) &= \cup \ \{C \in C_e \mid (e, C) \in Z\} \\ \alpha(r) &= \{v \in V \mid (r, v) \in Z\}. \end{aligned}$$

Then α satisfies the first and third condition. It satisfies the second condition of Definition 4.28 on Page 152 exactly when Z is round.

The selected $\alpha(e)$ and $\alpha(r)$ are not empty exactly when $Z \neq \emptyset$. [30] — That is — the graph G has an antipodality of p-range $\geqslant k$ exactly when Z is round in M and $Z \neq \emptyset$.

This proves the theorem. $\square$

4.3.2   Tilts and slopes

It is our aim to prove the following theorem.

Theorem 4.33. *Let G be a connected planar graph with $|V| \geqslant 2$. Let $p : E \to \mathbb{N}$ and let $k \in \mathbb{Z}^{\geqslant 0}$. Then G has p-carvingwidth at least k if and only if it has a vertex x with $p(\delta(x)) \geqslant k$ or an antipodality of p-range at least k.*

In this section we show 'only if'.

Let V be a set and let $\kappa : 2^V \to \mathbb{Z}$. [31]   The function κ is submodular if it satisfies for all $X, Y \in 2^V$

$$\kappa(X \cup Y) + \kappa(X \cap Y) \quad \leqslant \quad \kappa(X) + \kappa(Y).$$

For example, the function $\kappa(X) = p(\delta(X)) + c$ (for any constant c) is a submodular function $2^V \to \mathbb{Z}^{\geqslant 0}$. [32]

Let $\kappa : 2^V \to \mathbb{Z}$. Call a set X efficient if $\kappa(X) \leqslant 0$. Assume that κ satisfies

(a) $\kappa(X) = \kappa(V \setminus X)$ for all $X \subseteq V$

(b) κ is submodular

(c) all $X \subseteq V$ of cardinality one are efficient.

A bias is a collection $\mathcal{B}$ of efficient subsets of V such that

(B1) for each efficient set X exactly one of X and $V \setminus X$ is in $\mathcal{B}$

---

[30] Notice that when one of the $\alpha(e)$ and $\alpha(r)$ is empty then they are all empty. This follows from the second antipodality condition.

[31] In this section we slip and slide all the way down the rabbit hole;

bias $\to$ tilt $\to$
slope $\to$ antipodality

[32] The edges that are not counted on the left are those with one end in $X \setminus Y$ and the other in $Y \setminus X$.

(B2) $X, Y, Z \in \mathcal{B} \Rightarrow X \cup Y \cup Z \neq V$

(B3) $|X| = 1 \Rightarrow X \in \mathcal{B}$.

Robertson and Seymour proved the following lemma in their 'Graph Minors X.' [33]

[33] That is 'Graph Minors. X. Obstructions to tree-decomposition.' It appeared in: Journal of Combinatorial Theory, Series B, **52** (1991), pp. 153–190.

Lemma 4.34. *Exactly one of the following statements is true*

1. *there is a carving $\mathcal{C}$ such that $\kappa(X) \leqslant 0$ for all $X \in \mathcal{C}$*

2. *there is a bias $\mathcal{B}$.*

Proof. We only prove that not both statements are true.

Assume there is a carving $\mathcal{C}$ (as stated in the lemma) and a bias $\mathcal{B}$. We derive a contradiction as follows.

The carving $\mathcal{C}$ corresponds with a routing tree T. Let e be an edge of the routing tree. Let X and $V \setminus X$ be the leaves in the components of $\mathsf{T} - e$. Since all sets in $\mathcal{C}$ are efficient both X and $V \setminus X$ are efficient. Since $\mathcal{B}$ is a bias exactly one of X and $V \setminus X$ is in $\mathcal{B}$.

An incident pair (v, e) is a pair with $v \in V(\mathsf{T})$ and $e \in E(\mathsf{T})$ and $v \in e$. Let $X(v, e)$ be the set of leaves in the component of $\mathsf{T} - e$ that contains x. Call a pair (v, e) passive if $X(v, e) \notin \mathcal{B}$. By the above there are exactly $|E(\mathsf{T})|$ incident pairs that are passive.

There are $|E(\mathsf{T})| + 1$ vertices in T. It follows that there is a vertex $v \in V(\mathsf{T})$ such that for that all edges e that have v as an endpoint $X(v, e) \notin \mathcal{B}$.

Now assume that such a vertex v is a leaf. Then $|X(v, e)| = 1$ and since $\mathcal{B}$ is a bias $X(v, e) \in \mathcal{B}$ which contradicts the assumption.

Assume that v is not a leaf. Then there are three edges incident with v. The sets of leaves $X(v, e_i)$ satisfy

$$\bar{X}(v, e_1) \;\cup\; \bar{X}(v, e_2) \;\cup\; \bar{X}(v, e_3) \;=\; V$$
$$\text{where} \quad \bar{X}(v, e_i) \;=\; V \setminus X(v, e_i) \;\in\; \mathcal{B}.$$

— So — assuming that all three $X(v, e_i) \notin \mathcal{B}$ contradicts the assumption that $\mathcal{B}$ is a bias.

This shows a small part of the proof. FOR THE SAKE OF BREVITY we direct the reader to 'Graph Minors X' for the remainder of the proof. □

Definition 4.35. A <u>tilt</u> of order k is a collection $\mathcal{B}$ of subsets of V that satisfy $p(\delta(X)) \prec k$ and [34]

T1) Let $X \subseteq V$. Then $\mathcal{B}$ contains exactly one of X and $V \setminus X$ if and only if $p(\delta(X)) < k$.

T2) When $X, Y, Z \in \mathcal{B}$ then $X \cup Y \cup Z \neq V$.

T3) For all $x \in V$ $\{x\} \in \mathcal{B}$.

Corollary 4.36. *Let* G *be a graph with at least two vertices. Let* $p : E \to \mathbb{N}$ *and let* $k \in \mathbb{N}$ *be so that* $p(\delta(v)) < k$ *for all* $v \in V$. *Then* G *has* p-*carvingwidth* $\geqslant k$ *if and only if* G *has a tilt of order* k.

Proof. Let
$$\kappa(X) \quad = \quad p(\delta(X)) - k + 1.$$
Then $\kappa(X) \leqslant 0$ if and only if $p(\delta(X)) < k$. The statement follows from Lemma 4.34. □

LET US GET BACK TO THAT SURFACE Σ. Let G be embedded in Σ and let $k \in \mathbb{N}$. We show how the circuits in the graph define a slope. [35]

Definition 4.37. A <u>slope</u> of order $k/2$ is a function ins that assigns to every circuit C of length $< k$ exactly one of the two closed discs in Σ that have C as boundary. Furthermore, the function ins satisfies the following conditions.

(S1) Let C and C' be circuits of length $< k$ and assume that C is drawn within ins(C'). Then

$$\mathrm{ins}(C) \quad \subseteq \quad \mathrm{ins}(C').$$

[34] A 'tilt' is a sloping surface (one that makes you glide down). It's of course just a linear transformation of a bias.

[35] We call a 'circuit' the embedding of a cycle in Σ. Every circuit cuts the sphere in two discs. So, topologically speaking, every circuit is an 'equator.'

(S2) Let P_1, P_2 and P_3 be three paths of length $< k$ that run between two vertices u and v but that are otherwise vertex disjoint. Then

$$\mathrm{ins}(P_1 \cup P_2) \quad \cup \quad \mathrm{ins}(P_1 \cup P_3) \quad \cup \quad \mathrm{ins}(P_2 \cup P_3) \quad \neq \quad \Sigma$$

A slope is <u>uniform</u> if for every region $r \in R(G)$ there is a circuit C of length $< k$ such that

$$r \quad \subseteq \quad \mathrm{ins}(C).$$

Definition 4.38. Let $X \subseteq V$ be nonempty such that $G[X]$ is connected and $V \setminus X \neq \varnothing$ and $G[V \setminus X]$ connected. Then $\delta(X)$ is a <u>bond</u>.

Exercise 4.25

Show that the dual of a bond is the set of edges of a circuit in G^*.

Lemma 4.39. *Let G be connected and drawn on a sphere Σ. Let $p : E(G) \to \mathbb{N}$. Let $k \in \mathbb{N}$ be such that $p(\delta(x)) < k$ for all $x \in V(G)$. Define the graph G' as the graph obtained from the dual G^* by subdividing each edge $e^* \in E(G^*)$ $p(e) - 1$ times.* [36] *If G has a tilt of order k then G' has a uniform slope of order $k/2$.*

[36] A subdivision replaces an edge (in this case an edge of G^*) by a path of length 2. Subdividing ℓ times introduces ℓ new vertices in Σ on e which replace the edge by a path of length $\ell + 1$.

Proof. Let $\mathcal{B}$ be a tilt of order k in G. We define a slope in G' of order $k/2$ and show that it is uniform.

Let C be a circuit in G' of length $< k$ and let Δ_1 and Δ_2 be the two discs in Σ with boundary C. Notice that

$$p(\delta(V(G) \cap \Delta_i)) \quad = \quad |E(C)| \quad < \cdot \; k.$$

— So — exactly one of the sets $V(G) \cap \Delta_i$ is an element of $\mathcal{B}$ — say — $V(G) \cap \Delta_1 \in \mathcal{B}$. Define $\mathrm{ins}(C) = \Delta_1$. Then (by the tilt conditions T1 and T2) ins is a slope of order $k/2$ in G'. We show that ins is uniform.

Let $r \in R(G')$. Then $r \in R(G^*)$; say $r = v^*$ for some $v \in V(G)$. By T3: $\{v\} \in \mathcal{B}$. Choose $X \in \mathcal{B}$ maximal such that $v \in X$ and such that $G[X]$ is connected.

Let $Y = V(G) \setminus X$. By T2: $Y \neq \varnothing$. We show that $G[Y]$ is connected (and so $\delta(X)$ is a bond).

Let $Y_1, Y_2, \cdots$ be the components of $G[Y]$. Each $G[X \cup Y_i]$ is connected since G and $G[X]$ are connected. Also $\delta(Y_i) \subseteq \delta(X)$. Since X is maximal $X \cup Y_i \notin \mathcal{B}$. Notice that

$$\delta(X \cup Y_i) \quad \subseteq \quad \delta(X) \quad \Rightarrow \quad p(\delta(X \cup Y_i)) \quad < \quad k \quad \Rightarrow$$
$$Y \setminus Y_i \quad = \quad V(G) \setminus (X \cup Y_i) \quad \in \quad \mathcal{B}. \quad ((\text{By T1}))$$

When $t \geqslant 2$ then X, $Y \setminus Y_1$ and $Y \setminus Y_2$ are all in $\mathcal{B}$ and their union is $V(G)$. This contradicts T2.

— So — $t \leqslant 1$ and $t \neq 0$ (by T2). It follows that $\delta(X)$ is a bond. Let $C = \{e^* \mid e \in \delta(X)\}$. Then C is the set of edges of a circuit in G^*. The subdivisions transform C into a circuit C' of G'. Then

$$|E(C')| = p(\delta(X)) < k$$

and $r \subseteq \text{ins}(C') = \text{ins}(C)$ since $X \in \mathcal{B}$.

This proves that ins is a uniform slope of order $k/2$ in G'. $\qquad \square$

Robertson and Seymour proved the following lemma in their 'Graph Minors XI.' We omit the proof. [37]

[37] Graph Minors. XI. Circuits on a surface. Journal of Combinatorial Theory, Series B **60** (1994), pp. 72–106.

A closed walk W 'captures' a point $x \in \Sigma$ if it passes through x or there is a circuit C of length $< k$ that satisfies $E(C) \subseteq E(W)$ and $x \in \text{ins}(C)$.

Lemma 4.40. *Let G be drawn on a sphere Σ and let $k \in \mathbb{N}$. Let ins be a slope of order $k/2$ and let $x \in \Sigma$.*
Let $N_x \subseteq \Sigma$ be the set of $y \in \Sigma$ for which there is a closed walk W in G of length $< k$ that captures x and y. Then either $\Sigma - N_x$ is an open disc or $N_x = \varnothing$. Furthermore if ins is uniform then $N_x \neq \varnothing$.

WE COME TO THE FINAL STEP in showing that carvingswidth at least k and for all vertices $p(\delta(x)) < k$ implies an antipodality of p-range at least k. Actually, the following theorem shows that there is an antipodality which is connected. An antipodality α is <u>connected</u> if $\alpha(e)$ is connected for all $e \in E(G)$.

In the following theorem let G, G^*, G', p and k be as in Lemma 4.39.

Theorem 4.41. *If* G' *has a uniform slope of order* $k/2$ *then* G *has a connected antipodality of* p-*range* $\geqslant k$.

Proof. For $x \in \Sigma$ let N_x be as in Lemma 4.40 but for the graph G' instead of G.

Define α as follows. We show below that α is a connected antipodality of p-range at least k.

For $r \in R(G)$ let

$$\alpha(r) \quad = \quad \{v \in V \mid v^* \subseteq \Sigma - N_{r^*}\}$$

and for $e \in E(G)$ let

$$V(\alpha(e)) \quad = \quad \{v \in V \mid v^* \quad \subseteq \quad \Sigma - N_{x(e)}\}$$
$$E(\alpha(e)) \quad = \quad \{f \in E \mid f^* \quad \subseteq \quad \Sigma - N_{x(e)}\}.$$

Notice that this is a subgraph. [38]

By Lemma 4.40 $N_{x(e)}$ is an open disc — so — $\alpha(e)$ is a connected subgraph of G.

We show that α is an antipodality of p-range $\geqslant k$. To prove A1 let $e \in E$ and let v be an endpoint of e. Since ins is uniform there exists a circuit C in G' of length $< k$ such that $v^* \subseteq \text{ins}(C)$. Then $e^* \subseteq \text{ins}(C)$. So there is a circuit of length $< k$ that captures each point of v^* and $x(e)$ and this implies $v^* \subseteq N_{x(e)}$ — so — $v \notin V(\alpha(e))$.

To see A2 let $e \in E(G)$ and $r \in R(G)$ and let e and r be incident. Then e^* and r^* are incident in G^* and $N_{x(e)} \subseteq N_{r^*}$ (since any walk in G' that captures $x(e)$ also captures r^*) and so

$$\Sigma - N_{r^*} \quad \subseteq \quad \Sigma - N_{x(e)}.$$

[38] To see that let f be an edge with endpoint v. Notice that

$$f^* \subseteq \Sigma - N_{x(e)} \Rightarrow$$
$$v^* \subseteq \Sigma - N_{x(e)}$$

This show that $\alpha(r) \subseteq V(\alpha(e))$.

To see A3 let $e \in E$ and $f \in E(\alpha(e))$. No closed walk of length $< k$ captures both $x(e)$ an $x(f)$. This implies that no closed walk in G^* of length $< k$ contains e^* and f^*.

This proves the theorem. $\qquad \square$

THIS SHOWS THE FOLLOWING HALFWAY RESULT.

Let G be a connected graph with at least two vertices drawn on a sphere Σ with a dual G^*. Let $p : E(G) \to \mathbb{N}$ and $k \in \mathbb{N}$. Assume that $p(\delta(x)) < k$ for each vertex $x \in V(G)$ and that the p-carving width of G is $\geqslant k$. Then G has an antipodality of p-range $\geqslant k$.

4.3.3 Bond carvings

TAKE A DEEP BREATH: it's time to start the proof of the 'if' - part of Theorem 4.33 on Page 157.

LET'S GET IN THE MOOD with a groovy exercise. [39]

Exercise 4.26

Let $\mathcal{C}$ be a carving of a set V.

(i) $X \in \mathcal{C} \quad \Rightarrow \quad V \setminus X \in \mathcal{C}$

(ii) if $X \in \mathcal{C}$ and $|X| \geqslant 2$ then X has a unique partition $\{Y, Z\}$ with $Y, Z \in \mathcal{C}$. [40]

HINT: Use Exercise 4.20 on Page 150. A routing tree is ternary. The set X is the set of leaves of a branch and X has at least two leaves. (A branch of a tree is a component of the forest obtained by removing one edge of the tree.)

[39] The word 'groovy' is a bit outdated (it means 'cool' — if that's not already outdated also); it comes from the 'groove' in a vinyl record; 'the music is in the groove' ie in the carving of the record.

[40] So —

(a) $Y, Z \neq \varnothing$

(b) $Y \cap Z = \varnothing$

(c) $Y \cup Z = X$.

The fact that $Y, Z \in \mathcal{C}$ implies (also) that $Y, Z \neq \varnothing$.

Recall Definition 4.38 on Page 160: a set $\delta(X)$ is a bond if X and $V \setminus X$ induce connected subgraphs in G.

Definition 4.42. A carving $\mathcal{C}$ of a graph G is a <u>bond - carving</u> if $\delta(X)$ is a bond for all $X \in \mathcal{C}$.

Here's one more easy exercise.

Exercise 4.27

A carving $\mathcal{C}$ is a bond - carving if and only if X is connected for all $X \in \mathcal{C}$. [41]

For disjoint sets X and Y denote the set of edges that have one endpoint in X and the other in Y as $\delta(X, Y)$. Let $\mathcal{C}$ be a carving of G. A <u>triad</u> is a partition $\{X, Y, Z\}$ of V with $X, Y, Z \in \mathcal{C}$. — Clearly — each $X \in \mathcal{C}$ is in at most one triad. It is in a (unique) triad when

$$|X| \quad \leqslant \quad n - 2.$$

Define a 'measure' μ on the carvings of G as follows. Assume that G is connected and let $T = \{X, Y, Z\}$ be a triad. Then at most one of $\delta(X, Y)$, $\delta(X, Z)$, and $\delta(Y, Z)$ is empty. [42] If one is empty — say $\delta(X, Y) = \varnothing$ — then let $\mu(T) = |Z| - 1$. If none of $\delta(X, Y)$, $\delta(X, Z)$ and $\delta(Y, Z)$ is empty then let $\mu(T) = 0$. Define

$$\mu(\mathcal{C}) \quad = \quad \sum_{T \text{ a triad of } \mathcal{C}} \mu(T).$$

(The summation is over all triads $T = \{X, Y, Z\}$ in $\mathcal{C}$. [43])

Lemma 4.43. *Let $\mathcal{C}$ be a carving of G. Assume that G is connected and that $|V| \geqslant 2$. Let $\{A_1, A_2, B_1, B_2\} \subseteq \mathcal{C}$ be a partition of V such that* [44]

(a) $A_1 \cup A_2 \in \mathcal{C}$

(b) $\delta(A_1, A_2) = \varnothing$

(c) $\delta(A_1, B_1) \neq \varnothing$ *and* $\delta(A_2, B_2) \neq \varnothing$.

[41] **Abuse!**   We mean of course that $G[X]$ is connected when $\mathcal{C}$ is a bond carving and $X \in \mathcal{C}$.

[42] Otherwise, one of X, Y and Z is disconnected from the rest.

[43] That is — over all points of degree 3 in a routing tree.

[44] Take an edge e of a routing tree T such that both components of $T \setminus e$ have at least two leaves. Let $\{A_1, A_2\}$ and $\{B_1, B_2\}$ be a partition of the two sets of leaves of $T \setminus e$. The lemma rebuilds the tree (grouping $A_1 \cup B_1$ and $A_2 \cup B_2$ into sets of leaves) into one that has a smaller μ-value.

Define a carving $\mathcal{C}'$ *of* G *as follows*

$$\mathcal{C}' \;=\; (\mathcal{C} \;\setminus\; \{A_1 \cup A_2, B_1 \cup B_2\}) \;\cup\; \{A_1 \cup B_1, A_2 \cup B_2\}.$$

Then $\mu(\mathcal{C}') < \mu(\mathcal{C})$.

Proof.

$$\mu(\mathcal{C}) \qquad \mu(\mathcal{C}') \quad -$$
$$\mu(A_1 \cup A_2, B_1, B_2) \quad + \quad \mu(A_1, A_2, B_1 \cup B_2)$$
$$- \mu(A_1 \cup B_1, A_2, B_2) \quad - \quad \mu(A_1, B_1, A_2 \cup B_2).$$

Since $\{A_1, A_2, B_1 \cup B_2\}$ is a triad and $\delta(A_1, A_2) = \varnothing$ we have (by definition of μ)

$$\mu(A_1, A_2, B_1 \cup B_2) \quad = \quad |B_1 \cup B_2| - 1 \quad = \quad |B_1| + |B_2| - 1.$$

We prove the claim via the principle of contradiction; assume

$$\mu(A_1 \cup B_1, A_2, B_2) \quad + \quad \mu(A_1, B_1, A_2 \cup B_2) \quad \geqslant$$
$$\mu(A_1 \cup A_2, B_1, B_2) \quad + \quad |B_1 \cup B_2| \quad - \quad 1. \quad (4.7)$$

Since B_1 and B_2 are disjoint we conclude

$$(\,\mu(A_1 \cup B_1, A_2, B_2) \quad - \quad (\,|B_2| - 1\,)\,) \quad +$$
$$(\,\mu(A_1, B_1, A_2 \cup B_2) \quad - \quad (\,|B_1| - 1\,)\,) \quad > \quad 0.$$

— So — without loss of generality we may assume that

$$\mu(A_1, B_1, A_2 \cup B_2) \quad > \quad |B_1| \quad - \quad 1. \quad (4.8)$$

Notice that (4.8) implies that $\delta(A_1, A_2 \cup B_2) \neq \varnothing$ since otherwise the equation would be an equality. Since we are also given that $\delta(A_1, B_1) \neq \varnothing$ (4.8) implies that $\delta(B_1, A_2 \cup B_2) = \varnothing$ (since otherwise $\mu(A_1, B_1, A_2 \cup B_2) = 0$). — So — we obtain that

$$\mu(A_1, B_1, A_2 \cup B_2) \quad = \quad |A_1| - 1 \quad \text{and} \quad \delta(B_1, A_2 \cup B_2) = \varnothing.$$

— Clearly — this implies

$$\delta(B_1, B_2) \quad \subseteq \quad \delta(B_1, A_2 \cup B_2) \quad = \quad \varnothing \quad \Rightarrow$$
$$\mu(A_1 \cup A_2, B_1, B_2) \quad = \quad |A_1 \cup A_2| - 1.$$

Rewriting our assumption (4.7) we find that

$$\mu(A_1 \cup B_1, A_2, B_2) + |A_1| - 1 \;\;\geqslant\;\; |A_1 \cup A_2| - 1 + |B_1 \cup B_2| - 1 \;\;\Rightarrow$$
$$\mu(A_1 \cup B_1, A_2, B_2) \;\;\geqslant\;\; |A_2 \cup B_1 \cup B_2| - 1 \;\;>$$
$$\max\{|A_2| - 1, |B_2| - 1\}. \quad (4.9)$$

By assumption $\delta(A_2, B_2) \neq \varnothing$ — so — $\mu(A_1 \cup B_1, A_2, B_2) \neq |A_1 \cup B_1| - 1$. Then $\mu(A_1 \cup B_1, A_2, B_2)$ must be one of $|A_2| - 1$, $|B_2| - 1$, or 0. This contradicts (4.9).

This proves the lemma.   □

WE COME TO THE MAIN RESULT OF THIS SECTION. When G is a 2-connected graph and $p : E \to \mathbb{N}$ then G has a bond-carving of minimal p-width. [45]

[45] A graph is 2-connected if every minimal separator in it has at least two vertices.

Theorem 4.44. *Assume that a graph* G *is 2-connected and that it has at least two vertices. Let* $p : E \to \mathbb{N}$ *and assume that* G *has p-carving width* $< k$. *Then* G *has a bond - carving* $\mathcal{C}$ *such that* $p(\delta(X)) < k$ *for all* $X \in \mathcal{C}$.

Proof. SINCE WE INTRODUCED THAT MEASURE μ WE MIGHT AS WELL USE IT. Let $\mathcal{C}$ be a carving of G such that $p(\delta(X)) < k$ for all $X \in \mathcal{C}$ and assume that $\mu(\mathcal{C})$ is minimal (subject to the above). We claim that $\mathcal{C}$ is a bond-carving.

Assume not. Then there exists $X \in \mathcal{C}$ which is not connected. Choose X such that it has minimal size. (Clearly $|X| > 1$ since it induces a disconnected graph).

Since $\mathcal{C}$ is a carving there exists a partition $\{X_1, X_2\}$ of X with both sets $X_i \in \mathcal{C}$. By the minimality of X both X_1 and X_2 are connected. Since X is not connected $\delta(X_1, X_2) = 0$.

This shows that $\mathcal{C}$ has a triad $\{A_1, A_2, B\}$ such that

i. $\delta(A_1, A_2) = \varnothing$

ii. $|B|$ minimal subject to i.

The set B is a separator for A_1 and A_2. Since we assume that G is 2-connected $|B| \geqslant 2$. It follows that B has a partition $\{B_1, B_2\}$ with both $B_i \in \mathcal{C}$.

Claim **A**: $\delta(A_1, B_1) \neq \varnothing$ or $\delta(A_2, B_1) \neq \varnothing$. That is so because the triad $\{B_1, A_1 \cup A_2, B_2\}$ has $|B_2| < |B|$ and so $\delta(A_1 \cup A_2, B_1) \neq \varnothing$ (since we chose B minimal). For the same reason we have $\delta(A_1 \cup A_2, B_2) \neq \varnothing$.

First assume that $\delta(A_1, B_1) \neq \varnothing$ and $\delta(A_2, B_2) \neq \varnothing$. Claim: $p(\delta(A_1 \cup B_1)) \geqslant k$. To see that by Lemma 4.43 $\mu(\mathcal{C}') < \mu(\mathcal{C})$ and so (since $\mu(\mathcal{C}$ is minimal) there exists $X \in \mathcal{C}'$ such that $p(\delta(X)) \geqslant k$. By definition of $\mathcal{C}'$ either $X = A_1 \cup B_1$ or $X = A_2 \cup B_2$. In either case

$$p(\delta(A_1 \cup B_1)) \quad = \quad p(\delta(A_2 \cup B_2)) \quad = \quad p(\delta(X)) \quad \geqslant \quad k. \quad (4.10)$$

By similarity we also have that $\delta(A_1, B_2) \neq \varnothing$ and $\delta(A_2, B_1) \neq \varnothing$ implies

$$p(\delta(A_1 \cup B_2)) \geqslant k. \quad (4.11)$$

Since G is connected at least one of $\delta(A_1, B_1)$ and $\delta(A_1, B_2)$ is not empty and the same holds when A_2 replaces A_1. By Claim **A** we may assume that $\delta(A_1, B_1) \neq \varnothing$ and $\delta(A_2, B_2) \neq \varnothing$. By Equation (4.10) $p(\delta(A_1 \cup B_1)) \geqslant k$. Since $p(\delta(B_1)) < k$ and $\delta(A_1, A_2) = \varnothing$,

$$\delta(A_1 \cup B_1) \not\subseteq \delta(B_1) \quad \Rightarrow \quad \delta(A_1, B_2) \neq \varnothing,$$

and — similarly — $\delta(A_2, B_1) \neq \varnothing$. By (4.11) $p(\delta(A_1 \cup B_2)) \geqslant k$. We derive a contradiction as follows

$$
\begin{aligned}
2 \cdot k \quad \leqslant \quad & p(\delta(A_1 \cup B_1)) + p(\delta(A_1 \cup B_2)) \quad = \\
& p(\delta(A_1, B_2)) + p(\delta(A_2, B_1)) + p(\delta(B_1, B_2)) + \\
& p(\delta(A_1, B_1)) + p(\delta(A_2, B_2)) + p(\delta(B_1, B_2)) \quad = \\
& \qquad\qquad p(\delta(B_1)) + p(\delta(B_2)) \quad < \quad 2 \cdot k.
\end{aligned}
$$

(where the last line follows from $B_1, B_2 \in \mathcal{C}$)

This proves the theorem.   □

4.3.4   Carvings and antipodalities

In this section we complete the proof of Theorem 4.33 on Page 157. We present the proof in seven easy steps.

Let G be a connected planar graph with at least two vertices. Let G be drawn on a sphere Σ and let G^* be the dual. Let $p : E \to \mathbb{N}$ and let $k \in \mathbb{Z}^{\geq 0}$.

Let α be an antipodality of p-range $\geq k$. In this section we show that G has p-carving width $\geq k$. [46]

A pair (P, ν) with $P \subseteq V$ and $\nu \in P$ is a <u>limb</u> if

$$\delta(P) \subseteq \delta(\nu) \quad \text{and} \quad \exists_{e \text{ with endpoint } \nu} \quad V(\alpha(e)) \cap P \neq \varnothing.$$

Exercise 4.28

When (P, ν) is a limb and $P \neq V$ then ν is a cutvertex and P is the union of a collection of components of $G - \nu$ and $\{\nu\}$.
HINT: Use that $\delta(P) \subseteq \delta(\nu)$.

Lemma 4.45. *If* (P, ν) *is a limb then*

$$\forall_{e \text{ with endpoint } \nu} \quad V(\alpha(e)) \cap P \setminus \{\nu\} \neq \varnothing.$$

Proof. Let $e_1, \cdots, e_t$ be the edges that are incident with ν in the cyclic order in which they are drawn in Σ. Assume $V(\alpha(e_1)) \cap P \neq \varnothing$. We prove the lemma via contradiction: let $i > 1$ be minimal $V(\alpha(e_i)) \cap P = \varnothing$.

Then $V(\alpha(e_{i-1})) \cap P \neq \varnothing$. Let H be a component of $\alpha(e_{i-1})$ with $V(H) \cap P \neq \varnothing$. Then $\nu \notin V(H)$ since $\alpha(e_{i-1})$ does not contain an endpoint of e_{i-1} (since α is an antipodality).

Since $\delta(P) \subseteq \delta(\nu)$ and $\nu \notin V(H)$ and H is connected

$$V(H) \quad \subseteq \quad P.$$

Let $r \in R(G)$ be the region that is incident with e_{i-1} and e_i. By the second antipodality condition

$$V(H) \cap \alpha(r) \neq \varnothing \quad \text{and} \quad \alpha(r) \subseteq V(\alpha(e_i)).$$

[46] Recall the definition of an antipodality. For each region $\varnothing \neq \alpha(r) \subseteq V$ and for each edge $\alpha(e)$ is a subgraph (so it has at least one vertex) that does not contain an endpoint of e. The function α satisfies the following conditions.

(A1) for each edge $e \in E(G)$ the subgraph $\alpha(e)$ does <u>not</u> contain an endpoint of e

(A2) if an edge e and a region r are incident then $\alpha(r) \subseteq V(\alpha(e))$ and every component of $\alpha(e)$ has a vertex in $\alpha(r)$

(A3) if $e \in E$ and $f \in E(\alpha(e))$ then every closed walk in G^* that contains e^* and f^* has p-length at least k.

This implies

$$\varnothing \quad \neq \quad V(H) \cap \alpha(r) \quad \subseteq \quad P \cap \alpha(r) \quad \subseteq \quad P \cap V(\alpha(e_i)).$$

This contradicts the choice of i.

This proves the lemma. $\qquad\qquad\qquad\qquad\qquad\qquad\qquad\qquad$ $\square$

Notice that all vertices have degree at least 1 since G is connected and $|V| \geqslant 2$. It follows that for all vertices (V, v) is a limb. Choose a limb (P, v) such that $\underline{|P| \text{ is as small as possible}}$.

Lemma 4.46. $G[P \setminus \{v\}]$ *is connected.*

Proof. Notice that $P \setminus \{v\} \neq \varnothing$ since [47]

$$V(\alpha(e)) \cap (P \setminus \{v\}) \quad = \quad V(\alpha(e)) \cap P \quad \neq \quad \varnothing.$$

Suppose $P \setminus v$ is not connected. Then there exist subsets P_1 and P_2 satisfying

1. $P_1 \cup P_2 = P$

2. $P_1 \cap P_2 = \{v\}$

3. $\delta(P_1 \setminus v, P_2 \setminus v) = \varnothing$

4. $P_1 \setminus v \neq \varnothing$ and $P_2 \setminus v \neq \varnothing$.

Choose e incident with v such that $P \cap V(\alpha(e)) \neq \varnothing$. Then one of $P_1 \cap V(\alpha(e))$ and $P_2 \cap V(\alpha(e))$ is $\neq \varnothing$. — That is — (P_1, v) or (P_2, v) is a limb. This contradicts the choice of (P, v) as a limb with P minimal.

This proves the lemma. $\qquad\qquad\qquad\qquad\qquad\qquad\qquad\qquad$ $\square$

Exercise 4.29

Show that $P \setminus \{v\} \neq \varnothing$ and that $\delta(P \setminus \{v\}, v) \neq \varnothing$.

Let B be a maximal 2-connected subgraph of G that contains v and a neighbor of v in P. [48] Since $\delta(P) \subseteq \delta(v)$

$$V(B) \quad \subseteq \quad P.$$

[47] When (P, v) is a limb then $P \neq \{v\}$ since $P \cap V(\alpha(e)) \neq \varnothing$ for some e incident with v and $\alpha(e)$ does not contain v.

[48] An edge forms a 2-connected subgraph. So B contains at least v and a neighbor of v in P. Since B is 2-connected and $\delta(P) \subseteq \delta(v)$ B does not contain any vertex outside P; otherwise B would contain a cutvertex; namely v.

Lemma 4.47. *Every neighbor of* v *in* P *is in* $V(B)$:

$$N[v] \cap P \quad \subseteq \quad V(B).$$

Proof. By definition of B v has a neighbor $u_1 \in V(B)$. Assume v has another neighbor u_2 in P. Since $G[P \setminus \{v\}]$ is connected there is a circuit C in $G[P]$ that contains v, u_1, and u_2.

Then

$$|V(B) \cap C| \quad \geqslant \quad 2 \quad \Rightarrow \quad C \subseteq V(B).$$

This implies $u_2 \in V(B)$ and this proves the lemma.   □

Exercise 4.30

For $X \subseteq B$ show that there exists a unique set $\tilde{X}$ that satisfies

$$\tilde{X} \cap B = X \quad \text{and} \quad \delta(\tilde{X}) = \delta(X, B \setminus X).$$

HINT: Consider the union of components of $V \setminus B$ that have a neighbor in X. No component can have a neighbor in X and in $B \setminus X$ since B is a biconnected component.

Exercise 4.31

When $v \notin X$ then $\tilde{X} \subseteq P \setminus \{x\}$.

Exercise 4.32

Let a graph H be a subgraph of a graph G. Let $p : E \to \mathbb{N}$. Show that the p-carving width of G is at least the p-carving width of H. HINT: Let T be a routing tree for G. Remove leaves that are not in $V(H)$. Let T' be the result of this. How can we change T' into a routing tree for H?

ASSUME THAT B HAS p-CARVING WIDTH $< k$. We show that this assumption leads to a contradiction.

By Theorem 4.44 there exists a bond - carving $\mathcal{C}$ such that for all $X \subseteq B$

$$p(\delta(X, B \setminus X)) \quad = \quad p(\delta(\tilde{X})) \quad < \quad k.$$

Define

$$\mathcal{C}' \;=\; \{X \in \mathcal{C} \mid v \notin X \quad \text{and} \quad \exists_{e \in \delta(\tilde{X})} \; V(\alpha(e)) \cap \tilde{X} \neq \varnothing\}.$$

Lemma 4.48. $\mathcal{C}' \neq \varnothing$.

Proof. We show that $X = B \setminus \{v\} \in \mathcal{C}'$.

Clearly $X \in \mathcal{C}$. Notice that $\tilde{X} = P \setminus \{v\}$ (since every neighbor of v in P is in B). Choose $e \in E(B)$ with endpoint v. Since (P, v) is a limb (by Lemma 4.45)

$$V(\alpha(e)) \cap \tilde{X} \neq \varnothing.$$

Since $e \in \delta(\tilde{X})$ this proves that $X \in \mathcal{C}'$.   □

Choose $X \in \mathcal{C}'$ such that $|X|$ is minimal.

Lemma 4.49. $|X| \neq 1$.

Proof. Assume $X = \{u\}$. Then $\delta(\tilde{X}) \subseteq \delta(u)$ and $V(\alpha(e)) \cap \tilde{X} \neq \varnothing$ for some $e \in \delta(\tilde{X})$ (since $X \in \mathcal{C}'$). This shows that $(\tilde{X}, u)$ is a limb and $\tilde{X} \subseteq P \setminus v$. This contradicts the choice of (P, v) (as a limb with $|P|$ is as small as possible).   □

SINCE $\mathcal{C}$ IS A CARVING there exist $X_1, X_2 \in \mathcal{C}$ such that $\{X_1, X_2\}$ is a partition of X. By the minimality of $|X|$ neither X_1 nor X_2 is in $\mathcal{C}'$.

Lemma 4.50. *For all* $e \in \delta(\tilde{X}_1) \cup \delta(\tilde{X}_2)$

$$E(\alpha(e)) \cap \left(\delta(\tilde{X}_1) \cup \delta(\tilde{X}_2) \right) \;=\; \varnothing.$$

Proof. Let $e, f \in \delta(\tilde{X}_1) \cup \delta(\tilde{X}_2)$. One of $\delta(X)$, $\delta(X_1)$ and $\delta(X_2)$ contains both e and f — say D.

Since $\mathcal{C}$ is a bond-carving D is a bond and this implies that there is a circuit C in G^*

$$E(C) \quad = \quad \{ f^* \mid f \in D \}$$

and $e^*, f^* \in E(C)$.

The circuit C has p-length $p(D) < k$. By the antipodality condition A_3: $f \notin E(\alpha(e))$.

This proves the lemma. $\qquad\qquad\qquad\qquad\qquad\qquad\qquad\square$

Lemma 4.51. *For all $e \in \delta(\tilde{X})$*

$$V(\alpha(e)) \cap \tilde{X}_1 \quad = \quad \varnothing.$$

Proof. Since $\delta(X)$ is a bond we may choose an ordering of the edges in $\delta(\tilde{X})$

$$e_1 \quad \cdots \quad e_t$$

such that there is a region r_i in G incident with e_{i-1} and e_i.

Notice that $B \setminus X_2$ is connected. That is so because $B \setminus X_2 \in \mathcal{C}$ and $\mathcal{C}$ is a bond carving. It follows that

$$\delta(\tilde{X}_1) \cap \tilde{X} \quad \neq \quad \varnothing.$$

— So — we may choose $e_1 \in \delta(\tilde{X}_1)$. Since $X_1 \notin \mathcal{C}'$ we have

$$V(\alpha(e_1)) \cap \tilde{X}_1 \quad = \quad \varnothing.$$

Let i be minimal such that $V(\alpha(e_i)) \cap \tilde{X}_1 \neq \varnothing$. [49] Let H be a component of $\alpha(e_i)$ that intersects $\tilde{X}_1$. By Lemma 4.50 $E(H) \cap \delta(\tilde{X}_1) = \varnothing$. Since H is connected

$$V(H) \quad \subseteq \quad \tilde{X}_1.$$

[49] We go via contradiction.

By the second antipodality condition

$$V(H) \cap \alpha(r_i) \quad \neq \quad \varnothing$$

and $\alpha(r_i) \subseteq V(\alpha(e_{i-1}))$. This implies $V(\alpha(e_{i-1})) \cap \tilde{X}_1 \neq \varnothing$ and this is a contradiction. — So — $V(\alpha(e)) \cap \tilde{X}_1 = \varnothing$ for all $e \in \delta(\tilde{X})$.

The same argument shows that $V(\alpha(e)) \cap \tilde{X}_2 = \varnothing$ for all $e \in \delta(\tilde{X})$. Since $\{X_1, X_2\}$ is a partition of X $V(\alpha(e)) \cap \tilde{X} = \varnothing$ for all $e \in \delta(\tilde{X})$. This contradicts the assumption that $X \in \mathcal{C}'$.

This proves the lemma. □

IN OTHER WORDS the assumption on Page 170 is wrong: B has p-carving width $\geqslant k$. This shows that G has p-carving width $\geqslant k$ since B is a subgraph of G.

The following theorem summarizes the results.

Theorem 4.52. *Let G be a connected planar graph with at least two vertices. Let G be drawn on a sphere with a dual G^*. Let $p :$ $E \to \mathbb{N}$ and let $k \in \mathbb{Z}^{\geqslant 0}$. The following statements are equivalent.*

1. *the graph G has p-carving width at least k*

2. *the graph G has a tilt of p-order k*

3. *the graph G' has a uniform slope of order $k/2$* [50]

4. *the graph G has a connected antipodality of p-range at least k*

5. *the graph G has an antipodality of p-range at least k.*

[50] G' is obtained from G^* by subdividing edges $e^* \in E(G^*)$ $p(e) - 1$ times.

Remark 4.53. By Theorem 4.32 there exists an $O(m^2)$ - algorithm to decide if the p-carving width $\geqslant k$ ($m = |V| + |E|$). In their paper Robertson and Seymour show that the result above can be used to compute the 'branchwidth' of planar graphs in $O(m^2)$ time. The branchwidth parameter approximates the treewidth of graphs within a factor $3/2$.

4.4   Tree - degrees of graphs

CHORDAL GRAPHS ARE THE INTERSECTION GRAPHS OF SUB-TREES OF A TREE (see Exercise 4.5 on Page 133). In this section we look at graphs that have a — slightly — different intersection model.

Definition 4.54. A graph G is the <u>edge intersection - graph</u> of a tree T if there is a collection of subtrees

$$\{T_x \mid x \in V\}$$

such that

$$\{x, y\} \in E(G) \quad \Leftrightarrow \quad E(T_x) \cap E(T_y) \neq \varnothing.$$

THE DEFINITION ABOVE IS — A KIND OF — A JOKE. Namely any graph is the edge intersection - graph of a family of subtrees of a tree. To see that consider a star $K_{1,z}$ and a bijection from its leaves to the maximal cliques of G. [51] For $x \in V(G)$ let T_x be the subtree of the star that connects all leaves that contain x.

[51] So z is the number of maximal cliques in G. Two subtrees of a star share a line if and only if they share a leaf ie if and only if the vertices share a maximal clique.

Exercise 4.33

Show that the above constructs an edge intersection - model for any graph.

HINT: Two vertices are adjacent only if they are in a maximal clique — that is — only if their subtrees share a leaf.

Exercise 4.34

Let G be a graph without isolated vertices [52] and let $\mathcal{C}$ be a collection of maximal cliques that cover the edges of G — that is — every edge of G is contained in one of the cliques of $\mathcal{C}$. The

[52] If x is an isolated vertex let T_x be a subtree that contains only one (arbitrary) vertex.

minimal number of cliques in a cover of E is the edge clique - cover of G. Denote the edge - clique cover - number of a graph G as $cc(G)$. Its computation is NP-complete.

Say that G has an edge clique - cover with k cliques. Show that G is the edge intersection graph of a tree with maximal degree k.

Definition 4.55. The tree - degree $\tau(G)$ of a graph G is the minimal $k \in \mathbb{N} \cup \{0\}$ such that G is the edge intersection - graph of subtrees of a tree that has maximal degree $\leqslant k$.

Exercise 4.35

Let G be connected. Show that $\tau(G) = 1$ if and only if G is a clique with at least two vertices.

Assume that G has $k \geqslant 2$ components G_i that have at least one edge. Show that

$$\tau(G) \quad = \quad \max\{\tau(G_i), 2 \mid 1 \leqslant i \leqslant k\}$$

4.4.1 Intermezzo: Interval graphs

Consider a set of n intervals on the real line — say — $I_1, \cdots, I_n$. Construct a graph with vertex set $[n]$. Let two vertices be adjacent if the intervals have a nonempty intersection. The graph is called an interval graph.

Definition 4.56. A graph is an interval graph is it is the intersection graph of a collection of intervals on a line.

Exercise 4.36

Show that interval graphs are chordal.

A COLLECTION OF INTERVALS ON THE REAL LINE SATISFIES THE HELLY PROPERTY: if any pair of a collection of intervals intersects then they they all contain a common point on the real line. [53] Suppose we scan the line from left to right and for each point we record all the intervals that contain that point. That will give us a list of the cliques of G.

Exercise 4.37

Show that a graph G is an interval graph if and only if its maximal cliques can be put in a linear order [54]

$$C_1 \quad \cdots \quad C_t$$

such that for any vertex the cliques that contain it form an interval.

Exercise 4.38

Let G be an interval graph and let $C_1, \cdots, C_t$ be a consecutive clique arrangement. Show that the collection

$$\{ C_i \cap C_{i+1} \mid 1 \leqslant i < t \}$$

is the collection of minimal separators in G.

THE REASON FOR INTRODUCING INTERVAL GRAPHS is the following presentation of graphs that have tree - degree at most two and three.

Exercise 4.39

A graph satisfies $\tau \leqslant 2$ if and only if it is an interval graph.

A graph satisfies $\tau \leqslant 3$ if and only if it is a chordal graph.

---

[53] The Helly property also holds for subtrees of a tree: when $\mathcal{C}$ is a collection of subtrees of a tree T of which every pair intersects in some point of T. Then T contains a point that in in all the subtrees of $\mathcal{C}$. The same is not true when we consider edge - intersections: for example, consider three paths in a claw each connecting one pair of leaves.

[54] Since G is chordal it has at most n maximal cliques. The linear order of maximal cliques is called a consecutive clique arrangement.

Remark 4.57. It can be shown that for all graphs [55]

$$\tau(G) \quad \leqslant \quad cc(G).$$

Equality holds when G has no clique separator.

Let $k \in \mathbb{N}$ and let $\mathcal{G}$ be a class of graphs that satisfy $\tau \leqslant k$. There exists a polynomial - time algorithm to compute the treewidth of a graph in $\mathcal{G}$. The reason is that the number of minimal separators in a graph of $\mathcal{G}$ is at most $3m \cdot 2^{\tau-2}$. There exists a polynomial - time algorithm to compute the treewidth of graphs that have only a polynomial number of minimal separators. [56] [57]
An upper-bound for the treewidth is

$$tw \quad \leqslant \quad \tau \cdot \omega.$$

The computation of $\tau(G)$ remains NP-complete even when restricted to plane triangulations (G is a plane graph and every face is a triangle).

4.5 Modular decomposition

[55] M. Chang, T. Kloks and H. Müller, *On the tree-degree of graphs*, Springer, Lecture Notes in Computer Science **2204** (2001), pp. 44–54.

[56] V. Bouchitté and I. Todinca, *Minimal triangulations of graphs with "few" minimal separators*, Springer, Lecture Notes in Computer Science **1461** (1998), pp. 344–355.
[57] T. Kloks and D. Kratsch, Finding all minimal separators of a graph. Technical report 9327, Eindhoven University of Technology, 1993.

A module is a spacelab.

LET'S START WITH A DEFINITION; then we know what we are talking about.

Definition 4.58. Let G be a graph. A <u>module</u> in G is a set of vertices $X \subseteq V$ with the property that every vertex of $V \setminus X$ is either adjacent to every vertex of X or not adjacent to any vertex in X.

The components of a graph as well as the components of the complement are modules.

CLEARLY every graph has modules. The <u>trivial modules</u> are $\varnothing$, V and the subsets with one vertex. A graph is <u>prime</u> if it only has trivial modules.

Exercise 4.40

(a) Let G be a graph and let X be a module in G. A set $Y \subseteq X$ is a module in G if and only if Y is a module in $G[X]$.

(b) Let M be a module in a connected graph G and assume that M contains two vertices that are at distance > 2 in G. Then $M = V$.

   HINT: If $z \notin M$ then z is not adjacent to anything in M (otherwise every pair of vertices in M is at distance $\leqslant 2$). But G is connected — so — $M = V$.

(c) Let X be a module in G and let $z \in V \setminus X$. Show that X is a module in $G[N(z)]$ or in $G[V \setminus N[z]]$.

(d) The set of modules $\mathcal{M}$ of a graph is a <u>partitive family</u> — that is —

   - all trivial modules are in $\mathcal{M}$
   - when X and Y are modules that overlap then

$$X \cap Y \quad X \cup Y \quad X \setminus Y \quad \text{and} \quad (X \setminus Y) \cup (Y \setminus X)$$

   are all modules.

Two sets A and B <u>overlap</u> if $A \setminus B \neq \emptyset$ and $B \setminus A \neq \emptyset$ and $A \cap B \neq \emptyset$.

Definition 4.59. A module is <u>strong</u> if it does not overlap other modules.

Every graph has a partition of its vertices with parts that are strong modules.

Exercise 4.41

(a) Let A be a strong module. The smallest strong module that properly contains A is unique.

(b) Let S be a minimal separator in G and let M be a module in G that has a vertex of S but that is not contained in S. Then M contains all vertices of components that are close to S.

LET US SEE if a BFS - tree can be of any use in spotting a module.

A BFS - tree is a spanning tree with the property that every path from a vertex in the tree to the root is a shortest path in the graph.

In this chapter a BFS - tree T has a left - to - right order of the vertices in every level with the following property. If a vertex x has a parent y in T then x is not adjacent to any vertex in the level of y that is to the left of y.

Exercise 4.42

1. Let M be a module and let x be a vertex not in M. Let T be a BFS - tree with root x. Then all vertices of M have the same parent in T.

2. Let C be a component of the graph induced by the vertices in a level > 1 in a BFS - tree with root x. Let M be a strong module that contains x. Then $C \subset M$ or $C \cap M = \varnothing$.

3. Let x be a vertex and let C be a cocomponent in $G[N(x)]$. Let M be a strong module that contains x. Then $C \subset M$ or $C \cap M = \varnothing$.

4.5.1 Modular decomposition tree

The nodes of a modular decomposition tree are the nonempty modules that overlap with no other modules. The ancestor relation is containment.

LET G BE A GRAPH. A modular decomposition - tree for the graph G is a rooted tree with the vertices of G as leaves defined as follows.

Exercise 4.43

Show that a decomposition tree (as defined below) has the following property. The leaves of a branch are a module in G.

1. if G is disconnected then the root is labeled as a <u>parallel node</u>. Its children are the roots of the decompositions of the components.

2. if $\bar{G}$ is disconnected then the root is labeled as a <u>series node</u>. Its children are the roots of the decomposition trees of the cocomponents.

3. if G and $\bar{G}$ are both connected then there is a set $U \subseteq V$ and a partition $\mathcal{P}$ of V with the following properties.

- $|U| > 3$
- $G[U]$ is a maximal prime subgraph of G
- every part $P \in \mathcal{P}$ satisfies $|U \cap P| = 1$.

The node is a <u>prime node</u> and labeled as $G[U]$. Its children are the roots of a decomposition tree for the parts in $\mathcal{P}$.

Every graph with less than 4 vertices is a cograph. The decomposition tree for a cograph has only series nodes and parallel nodes.

Can two adjacent nodes in a modular decomposition tree be of the same type?

Exercise 4.44

Let G be a cograph. What are the strong modules in G?

HINT: The cotree is a modular decomposition (without any prime nodes). Every branch in a cotree is a module. Every induced subgraph has a module with two vertices.

4.5.2 A linear - time modular decomposition

In this section we take a look at the following paper.

M. Tedder, D. Corneil, M. Habib and C. Paul, Simple, linear - time modular decomposition. Manuscript on arXiv: 07010.3901, 2008.

The algorithm that is presented in this paper computes a modular decomposition tree as follows.

We assume that G is connected. Choose an arbitrary vertex x and let $\{L_i\}$ be a partition of V into the levels of a BFS-tree with $L_0 = \{x\}$ and $L_1 = N(x)$ and so on. The algorithm recursively computes a modular decomposition tree $T(L_i)$ for each layer L_1 and orders them as

$$T(L_1) \quad \{x\} \quad T(L_2) \quad T(L_3) \quad \cdots$$

Recall: The strong modules that do not contain x are modules in $N(x)$ or in $V \setminus N[x]$. The difficulties are

1. find the strong modules that contain x

2. construct the modular decomposition tree.

The aim is to compute a linear order of the vertices that is factorizing — that is — all the vertices of any strong module are consecutive.

We describe three procedures to achieve this. As an invariant we have an ordered partition of the vertices (each part is a set of leaves in a tree) which satisfies the condition that all the strong modules that do not contain x appear in consecutive parts.

> The invariant is a relaxation of the desired result.
>
> Below we add a second invariant that describes the positions of the strong modules that contain x.

Refinement

Let the current list of trees be

$$T(N_0) \quad \{x\} \quad T(N_1) \quad \cdots \quad T(N_k)$$

> The paper of Tedder et al uses
>
> $$N_0 = N(x) = L_1.$$

Let M be a strong module that does not contain x. Then $M \subseteq N_i$ for some i — furthermore — it is either a node in $T(N_i)$ or it is a set of nodes that all have the same parent.

The refinement - operation rebuilds $T(N_i)$ in case M is not a node. Let y be the parent of elements of M and assume that y has some children that are not in M. The procedure creates a new child y' of y. The children of y that are in M become children of y'.

Exercise 4.45

After a refinement the strong modules that do not contain x appear in consecutive parts.

When the refinement procedure terminates the nodes that do not have marked children are the strong modules containing x. The next procedure creates a new list of trees. This needs to be in order so that the invariant remains true. A "marker" helps us decide upon the new order.

Definition 4.60. An edge $\{x, y\} \in E(G)$ is active if its endpoints are in different N_i's.

Exercise 4.46

Upon termination of the refinement - procedure the strong modules that contain x are exactly the nodes that do not have marked children.

---

input an ordered list of trees:

$$T(N_0) \quad \{x\} \quad T(N_1) \quad \cdots \quad T(N_k)$$

$\alpha(v) = \{\, y \in N(v) \mid \{v, y\} \text{ is active}\,\};$
Refine with $X = \alpha(v)$ and
if v is left of x **OR** refine operates on a tree left of x **then**
    Refine with **left - splits** and marker=**left**
else
    Refine with **right - splits** and marker=**right**
end if

Algorithm 6: Part 1 of **refine**: decide on left-split, right-split, and marker

The presentation of the refinement - code is unorthodox. The first part shows what markers and splits are used. The details of the refinement procedure are in the second part.

A SECOND INVARIANT limits the trees in the sequence that contain vertices of strong modules that contain x. This invariant is formulated as follows.

Let the ordered set of trees — after a refinement — be

$$T_k \quad \cdots \quad T_1 \quad \{x\} \quad T'_1 \quad \cdots T'_\ell.$$

Let M be a strong module with $x \in M$. There exist trees T_i and T'_j such that

$$M \supset T_{i-1} \cup \cdots \cup T_1 \cup \{x\} \cup T'_1 \cup \cdots \cup T'_{j-1}$$
$$\text{and} \quad M \subseteq T_i \cup \cdots \cup T_1 \cup \{x\} \cup T'_1 \cup \cdots \cup T'_j.$$

Remark 4.61. If $M \neq V$ is a module and $x \in M$ then M has only vertices in levels 0, 1 and 2. If some vertex in level 2 is not in M then it is not adjacent to any vertex of M. So M intersects the second level in some union of components. (The same must hold

Let $T_1 \cdots T_k$ be the maximal subtrees with all leaves in X;
Let $P_1 \cdots P_\ell$ be the parents of $T_1, \cdots, T_k$;

for all P_i NOT prime **do**
    Partition the children of P_i ;
    A are the children of P_i that are T_j's and B are the other children;
        Create trees T_a and T_b with roots that have children A and B;
        Assign the label (series, parallel or prime) of P_i to A and B;
        if P_i is a root **then**
            if left - split **then**
                replace P_i with T_a T_b
            else
                replace P_i with T_b T_a
            end if;
        else
            replace children of P_i with T_a and T_b
        end if;
        Use the marker (left or right) to mark roots of T_a and T_b and all their ancestors ▷ (When a node is not a module then neither is any of its ancestors.)
end for;
for all P_i prime **do**
    Mark P_i; Mark all its children; Mark all its ancestors
end for

for $\bar{G}$ — so — M intersects $N(x)$ in a union of cocomponents.)
When the second level is disconnected then the root of $T(L_2)$ is a
parallel node and M is a set of children of the root.

Exercise 4.47

Let M be a strong module that contains x and a vertex at distance
> 2 from x. Then $M = V$. This shows that the second invariant
is true for the initial set of trees

$$T(N_0) \quad \{x\} \quad T(N_1) \quad \cdots \quad T(N_k)$$

with $T_i = T(N_0)$ and T_j either $T(N_1)$ or $T(N_k)$. We leave it as an
exercise to check that this invariant remains true after refinement.

Promotion

---

while there is a root r and child c both marked left **do**
    Move the branch with root c to the left of r
end while;
while there is a root r and child c both marked right **do**
    Move the branch with root c to the right of r
end while;
If a marked root has only one child then replace it with that child;
Delete all marked roots that have no children;
Remove all marks

Algorithm 8: Promotion

Exercise 4.48

Upon termination of the promotion - procedure the ordered list of
trees is a factorizing permutation.

Assembly

AT THIS POINT we have an ordered list of trees. The nodes of these trees (except x) are the strong modules that do not contain x and each of these is properly decomposed. What's left to do is to identify the strong modules that contain x.

The strong modules that contain x are nested in intervals around x. The (co-)components of $G[N_i]$ appear consecutive. The list of trees is a list of (co-)components

$$C_\kappa \quad \cdots \quad C_1 \quad \{x\} \quad C'_1 \quad \cdots \quad C_\lambda$$

where $\{C_i\}$ are the cocomponents of $G[N_0]$ and $\{C'_i\}$ are the components of $G[N_i]$ for $i > 0$. The strong components that contain x form a nested family of intervals.

By Exercise 4.42 we have the following lemma.

Lemma 4.62. *Let* M *be the smallest strong module that contains* x. *Then* M *satisfies one of the following.*

1. M *is a series - module:* M *is a maximal consecutive collection that contains* x *and no* C'_i

2. M *is parallel:* M *is a consecutive collection that contains* x *and no* C_i. *Furthermore all the* C'_j *that are in* M *are in* N_1 *and they have no edge to their right*

3. M *is prime:* M *is a consecutive collection which includes* $\{x\}$, C_1 *and* C'_1.

The lemma above is used to compute the strong modules that contain x as follows.

1. For C'_i in N_1 determine if it has a vertex with a neighbor in N_j for $j > 1$.

2. Determine a μ - value for each C_i and C'_i as follows.

- to find $\mu(C_i)$: find C'_j with smallest j such that C_i has no neighbors in C'_ℓ for $\ell \geqslant j$. then

$$\mu(C_i) \quad = \quad \begin{cases} x & \text{if } j = 1 \\ C'_{j-1} & \text{otherwise} \end{cases}$$

- $\mu(C'_i)$ is defined "symmetrically."

3. The cases in the lemma above are now easily recognized. If there is no series or prime module then M contains C_1 and C'_1. When C_i is added to M — recursively — add $C'_1 \cup \cdots \cup \mu(C_i)$ (and a symmetric rule applies when C'_j is added).

Conclusion

Theorem 4.63. *There exists a linear time algorithm to compute a modular decomposition tree of a graph.*

Exercise 4.49

LET'S DO IT AGAIN: Design an algorithm to compute a modular decomposition via a depth - first search tree.

Further reading

T. Uno and M. Yagiura, Fast algorithms to enumerate all common intervals of two permutations, *Algorithmica* **26** (2000), pp. 290 – 309.

4.5.3   Exercise

Let $\mathcal{G}$ be a class of graph. A graph G is a <u>probe graph</u> of $\mathcal{G}$ if its set of vertices partitions into probes and nonprobes such that G embeds in a graph of $\mathcal{G}$ by adding edges between nonprobes.

Exercise 4.50

Design an algorithm to check if a graph is a probe permutation graph. You may assume that a partition of the vertices into probes and nonprobes is a part of the input.

Hint: A graph G is a permutation graphs if and only if G and $\bar{G}$ are both comparability graphs. There is a linear - time algorithm to construct a transitive orientation of a comparability graph via modular decomposition. This does <u>not</u> imply a linear - time recognition algorithm for comparability graphs.

The modules in a permutation graphs can be represented by boxes in the permutation graph diagram. The following paper addresses the problem in detail.

D. Chandler, Maw-Shang Chang, T. Kloks, J. Liu, Sheng-Lung Peng, On probe permutation graphs, *Discrete Applied Mathematics* **157** (2009), pp. 2611–2619.

4.6   Rankwidth

Let G be a graph and let $\mathcal{C}$ be a carving of G. The <u>cut matrix</u> of a set $X \in \mathcal{C}$ is the submatrix of the adjacency matrix with X as rows and $V \setminus X$ as columns. Let $\mathsf{rank}(X)$ be the rank over $\mathrm{GF}[2]$ of the cutmatrix associated with X. [58] [59]

Definition 4.64. A graph has <u>rankwidth</u> $\leqslant k$ if it has a carving $\mathcal{C}$ such that for every $X \in \mathcal{C}$ $\mathsf{rank}(X) \leqslant k$.

[58] The rows of a cutmatrix form elements of a vector space over $\mathrm{GF}[2]$; the Galois field with two elements 0 and 1. (BTW Évariste Galois was a French mathematician.) The rank of a cutmatrix is the maximal number of rows of which no nonempty subset adds up to 0. (Addition of vectors is element-wise and obeys $0 + 0 = 0$, $1 + 0 = 1$, and $1 + 1 = 0$).

[59] The routing tree is restricted so that internal nodes have degree 3. This is of importance; a star would allow a decomposition of rankwidth 1 of any graph.

Exercise 4.51

Let $\mathcal{C}$ be a carving of a graph G of rankwidth $\leqslant k$. Show that every $X \in \mathcal{C}$ has a partition into at most 2^k classes such that the vertices of each class all have the same neighbors in $V \setminus X$.

Hliněný and Oum showed (in 2008) that there is a fixed-parameter $O(n^3)$ algorithm that recognizes graphs of rankwidth at most k (for $k \in \mathbb{N}$). (Computing the rankwidth of a graph is NP-complete.) Hliněný and Oum's algorithm finds a carving of rankwidth $\leqslant k$ if there exists one. [60]

[60] P. Hliněný and S. Oum, *Finding branch-decompositions and rank-decompositions*, SIAM Journal on Computing **38** (2008), pp. 1012–1032.

Geelen, Kwon, McCarthy and Wollan show that any circle graph H is a vertex - minor of a graph with sufficiently large rankwidth.

J. Geelen, O. Kwon, R. McCarthy and P. Wollan, The grid theorem for vertex - minors. Manuscript on arXiv: 1909.08113, 2019.

4.6.1 Distance hereditary - graphs

IT MAKES SENSE to have a close look at graphs that have rankwidth at most one. In this section we characterize those graphs.

Definition 4.65. A graph G is <u>distance - hereditary</u> if for any two nonadjacent vertices x and y all chordless $x \sim y$-paths have the same length. [61]

[61] A path is <u>chordless</u> if only consecutive pairs of vertices are adjacent in the graph.

PIONEERING WORK ON DISTANCE HEREDITARY - GRAPHS WAS DONE BY HOWORKA. We summarize some results below. [62]

[62] Are you getting confused? IF ALL ELSE FAILS you can always have a look at: Ton Kloks and Yue-Li Wang's <u>Advances in graph algorithms</u>.

Exercise 4.52

1. Show that a distance hereditary - graph G does not have an induces subgraph which is isomorphic to a house, hole, domino or gem. [63]

[63] A domino is a tile in a game and the game is called 'dominoes.' It seems that the game originated in China. (A domino is not a pizza!) The domino is a rectangle that is partitioned into two squares. Each square has a number of dots on it. For plural we use 'dominoes'.

2. Show that a graph is distance - hereditary if and only if it does
   not contain a house, hole, domino or gem as an induced subgraph.
   64

HINT: The class of graphs that are distance - hereditary is hereditary
— that is — the class is closed under taking induced subgraphs.
The house, holes, domino and gem are exactly the smallest graphs
that have two nonadjacent vertices that are connected by two chordless paths of different length (so the two nonadjacent vertices are
in a cycle).

64
(i) a <u>hole</u> in a graph is a chordless cycle of length at least 5

(ii) a <u>house</u> is C_5 with one chord

(iii) a <u>gem</u> is C_5 with two chords that share an end-point

(iv) a <u>domino</u> is C_6 with one chord that connects two vertices that are at distance 3 in C_6.

Exercise 4.53

Show that the class of graphs that are distance - hereditary is closed
under the following operations.

1. add a pendant vertex to the graph — that is — add one new
   vertex and give it exactly one neighbor

2. add a (true or false) twin — that is — add one new vertex
   and give it exactly the same (closed or open) neighborhood as
   one other vertex.

Exercise 4.54

Let T be a routing tree for a graph G of rankwidth $k \geqslant 1$. Let the
graph G′ be obtained from G by adding one new vertex that gets
the same (open or closed) neighborhood as a vertex of G. (That is;
G′ is obtained from G 'by creating a twin.') Construct a routing
tree for G′ (of width k).

One other (similar) question: how do you construct a routing tree
for the graph obtained from G by adding a pendant vertex (adjacent
to exactly one other vertex in the graph)?

Exercise 4.55

When G is distance - hereditary then every neighborhood induces
a cograph. 65

65 Recall Chapter 2.11: a
graph is a cograph if it does
not contain a path with 4 vertices as an induced subgraph.

HINT: Assume G has an induced P_4 with all its points in a neighborhood — say — $N(x)$ contains an induced P_4. Then G has a gem.

Theorem 4.66. *A graph has rankwidth $\leqslant 1$ if and only if it is distance - hereditary.*

Proof. Assume a graph G is distance - hereditary. By Exercise 4.53 G has a routing tree with the following property. The set of leaves — say X — of a branch has a partition $\{X_1, X_2\}$ (where we allow parts to be empty) such that the vertices of each part have the same neighbors in $V \setminus X$. Furthermore when both parts are nonempty then the vertices of one of the two parts have no neighbors in $V \setminus X$. [66] — In other words — all nonzero rows of the cutmatrix of X are the same. This shows that G has rankwidth $\leqslant 1$.

A carving $\mathcal{C}$ of rankwidth $\leqslant 1$ has the property that $\mathsf{rank}(X) \leqslant 1$ for every $X \in \mathcal{C}$. This implies that all vertices of X that have neighbors in $V \setminus X$ have the same neighbors in $V \setminus X$.

Let $X \in \mathcal{C}$ with $|X| = 2$. If X has two vertices that have neighbors in $V \setminus X$ then those two are twins. If X has only one vertex with neighbors in $V \setminus X$ then the other one is a pendant or an isolated vertex. [67] By induction it follows that G is distance - hereditary.

This proves the theorem. $\qquad\qquad\qquad\qquad\qquad\qquad\qquad\qquad$ $\square$

[66] A suitable routing tree is easily constructed via an elimination by pendant vertices and elements of twins.

[67] Clearly the class of distance hereditary - graphs is (also) closed under adding isolated vertices.

4.6.2 Intermezzo: Perfect graphs

The classes of chordal graphs, bipartite graphs and distance hereditary - graphs share an interesting property — namely — they are perfect.

Definition 4.67. A graph G is <u>perfect</u> if every induced subgraph H satisfies

$$\chi(H) \quad = \quad \omega(H).$$

THERE ARE TWO IMPORTANT THINGS TO SAY ABOUT PERFECT GRAPHS. They are called 'the perfect graph theorem' and 'the strong perfect graph theorem.' We present them without proof.

Theorem 4.68. *A graph is perfect if and only if its complement is perfect.*

Theorem 4.69. *A graph is perfect if and only if it does not contain an odd hole or an odd antihole as an induced subgraph.* [68]

Remark 4.70. A FEW OTHER THINGS: Claude Berge introduced perfect graphs after listening to a lecture by Claude Shannon. [69] In 1963 he proposed two conjectures that are now the two theorems above. [70]

 Important classes of perfect graphs are bipartite graphs, line graphs of bipartite graphs, chordal graphs and comparability graphs (and of course the classes of complements of these graphs).

One other characterization states that a graph is perfect if every induced subgraph H satisfies [71]

$$\alpha(H) \cdot \omega(H) \quad \geqslant \quad |V(H)|.$$

Lovász proved the perfect graph theorem in 1972. Chudnovsky, Cornuéjols, Liu, Seymour and Vušković showed (in 2008) that there is a polynomial - time algorithm to check if a graph is perfect.

Grötschel, Lovász, and Schrijver showed (in 1988) that $\alpha(G)$, $\omega(G)$, and $\chi(G)$ are computable in polynomial time on perfect graphs.

4.6.3 χ - Boundedness

As we already mentioned problems that can be formulated in monadic second-order logic (without using quantification over subsets of edges) can be solved in $O(n^3)$ time on graphs of bounded

[68] An odd hole in a graph is an induced cycle of odd length at least 5. An odd antihole in a graph is an odd hole in the complement.

[69] A lecture about the Shannon capacity of a graph. See: C. Shannon, *The zero-error capacity of a noisy channel*, IRE Trans. Inform. Theory (1956), 8–19.

[70] A graph is Berge if it does not contain an odd hole or odd antihole.

[71] Notice that this is not true for odd cycles of length more than 3.

rankwidth. When a routing tree is a part of the input then this reduces to $O(n)$ time.

Exercise 4.56

Let $k \in \mathbb{N}$. Design a monadic second-order formula that checks if a graph G can be (properly) colored with at most k colors — that is — the formula expresses $\chi(G) \leqslant k$.

HINT: A graph is k-colorable if there exists a partition of $V(G)$ into at most k classes that are all independent sets in G.

— THE IDEA IS — to prove that χ is computable in $O(n^3)$ time for graphs of bounded rankwidth and bounded clique number by providing an upper bound

$$\chi \quad \leqslant \quad f(\omega) \tag{4.12}$$

for some function $f : \mathbb{N} \to \mathbb{N}$ such that (4.12) holds true for all graphs of rankwidth $\leqslant k$.

Definition 4.71. A class of graphs $\mathcal{G}$ is $\underline{\chi\text{-bounded}}$ if there exists a function $f : \mathbb{N} \to \mathbb{N}$ such that

$$\chi(G) \quad \leqslant \quad f(\omega(G))$$

for all $G \in \mathcal{G}$.

In this section we show that for $k \in \mathbb{N}$ the class of graphs of rankwidth $\leqslant k$ is χ-bounded.

Many classes of graphs are χ-bounded; for example intersection graphs of axis-parallel boxes in d-space; graphs without odd holes; graphs without long holes; graphs that do no contain a subdivision of a tree. *Circle graphs* are $\underline{\text{polynomially}}$ χ-bounded.

Z. Dvořák and D. Král', *Classes of graphs with small rank decompositions are χ-bounded*, Manuscript on arXiv: 1107.2161, 2011.

Exercise 4.57

Show that the class of distance hereditary - graphs is χ-bounded.
HINT: Design an algorithm that colors a distance hereditary - graph with ω colors.

We show   that there is a partition of V that splits up all maximum cliques.

Lemma 4.72. *Let* G *be a connected graph with at least two vertices and of rankwidth* $\leqslant k$. *There exists a partition of* $V(G)$ *into at most* $3 \cdot 2^k$ *classes such that each class induces a subgraph with clique number less than* $\omega(G)$.

Proof. We may assume that the graph G has no false twins; ie the graph has no two vertices x and y with $N(x) = N(y)$. [72]

Let T be a routing tree. To facilitate the description of the partitioning - procedure we give T a root r which is a leaf and not a vertex of G. [73]

The tree T is ternary so we can label the lines of T with labels from $\{1, 2, 3\}$ such that each internal point of T is incident with an edge of each label. [74]

For $v \in V(T)$ let V_v denote the set of vertices of G that are leaves of the subtree T_v rooted at v. Since T has rankwidth k we can fix a partition

$$V_v \;=\; \{\, V_v^0, \;\; V_v^1, \;\; \cdots, \;\; V_v^d \,\}$$

such that

(a) the vertices of V_v^0 have no neighbors outside V_v

(b) all vertices of V_v^j for $j > 0$ have the same neighbors outside V_v

(c) $d < 2^k$.

Define a coloring ϕ of the vertices of G as follows. [75] To color $x \in V(G)$ find the point ℓ on the path from x to r in T that is furthest from r such that $N[x]$ is contained in the set of leaves of T_ℓ. — In other words V_ℓ is the minimal element of the carving that contains $N[x]$. By definition $x \in V_\ell^0$. Assume that x is a leaf of T_p for a child p of ℓ. Let $\alpha \in \{1, 2, 3\}$ be the label of the edge $\{\ell, p\} \in E(T)$.

In the partition of V_p find β such that $x \in V_p^\beta$. Notice that by the choice of ℓ [76]

$$0 \quad < \quad \beta \quad < \quad 2^k.$$

Color the vertex x with a pair:

$$\phi'(x) \quad = \quad (\alpha, \beta).$$

Map the colors that are used by ϕ' to a set of colors that is not used by ϕ to color $V_\ell^1, V_\ell^2, \cdots$. (Each of these classes is monochromatic.) This defines ϕ.

This procedure describes a coloring of $V(G)$ that uses less than $3 \cdot 2^k$ colors. This coloring is not necessarily proper — however — it colors no maximum clique monochromatic.

To SEE THAT let K be a clique in G of size $\omega(G)$ and assume that ϕ colors K monochromatic. Then there exists a vertex $z \in V(T)$ such that

$$K \quad \subseteq \quad V_z^j$$

for some $j > 0$. [77] This implies that the vertices of K have a common neighbor in $V \setminus V_z$. This contradicts that K has size $\omega(G)$.

[77] EXERCISE !

This proves the lemma. □

THE REST IS A WALK IN THE PARK.

Theorem 4.73. *Let* $k \in \mathbb{N}$. *The class of graphs of rankwidth* $\leqslant k$ *is* χ-*bounded.*

Proof. Define a function $f : \mathbb{N} \to \mathbb{N}$

$$\text{for } s \in \mathbb{N}: \quad f(s) \quad = \quad 2^{k \cdot s} \cdot 3^{s-1}$$

Let G be a graph of rankwidth $\leqslant k$. We show that $\chi \leqslant f(\omega)$ by induction on ω.

When $\omega = 1$ then $\chi = 1 \leqslant f(1)$.

When $\omega > 1$ then by Lemma 4.72 there is a coloring ϕ of G that uses at most $3 \cdot 2^k$ colors and has the property no maximum clique is monochromatic.

The subgraphs induced by the color classes of ϕ have rankwidth $\leqslant k$ — so — each color class of ϕ has a proper coloring with at most $f(\omega - 1)$ colors.

Color a vertex x by a pair:

(1) the index of the color class of ϕ that contains it

(2) the color assigned by the proper coloring.

The number of colors used by this proper coloring of G is at most

$$3 \cdot 2^k \cdot f(\omega - 1) = f(\omega)$$

This proves the theorem. $\qquad\qquad\square$

H. Guo, T. Kloks, H. Wang and M. Xiao, *On conflict - free colorings of some classes of graphs.* Manuscript 2019.

Exercise 4.58

A <u>conflict - free</u> coloring of a graph is a function $c : V \to \mathbb{N}$ which satisfies the following condition

$$\forall_{x \in V} \; \exists_{y \in N[x]} \; \forall_{z \in N[x] \setminus \{y\}} \quad c(y) \;\neq\; c(z). \qquad (4.13)$$

Show that there is an $O(n^3)$ algorithm to compute the conflict-free chromatic number $\chi_{CF}(G)$ for graphs of bounded rankwidth and bounded clique number.

HINT: CLEARLY $\chi_{CF}(G) \leqslant \chi(G)$. A class of graphs of rankwidth $\leqslant k$ is χ-bounded. This solves the problem since (4.13) is a formula in monadic second-order logic.
The real question is to find a good upper bound for

$$\max \{ \chi_{CF}(G) \mid \text{rankwidth}(G) \leqslant k \}.$$

Remark 4.74. Gyárfás did a lot of pioneering work on χ-bounded classes of graphs. [78] One conjecture of his is that for any tree T the class of graphs that do not have T as an induced subgraph is χ-bounded. A weaker statement is true — namely — for every tree T the class of graphs that do not contain any subdivision of T as an induced subgraph is χ-bounded.

By Erdős' result the class of triangle-free graphs is *not* χ-bounded. Scott and Seymour proved (in 2016) that, for all $\kappa \geqslant 0$, if G is a a graph with

$$\omega \leqslant \kappa \quad \text{and} \quad \chi > 2^{2^{\kappa+2}}$$

then G has an odd hole.

For more general information on χ-boundedness we direct the reader to the survey by Scott and Seymour. [79]

[78] A. Gyárfás, *Problems from the world surrounding perfect graphs*, Proceedings of the international conference on combinatorial analysis and its applications (Pokrzywna, 1985), Zastos. Mat. **19** (1987), pp. 413–441.

To pronounce Gyárfás, try to say something like "garfas."

[79] A. Scott and P. Seymour, *A survey of χ-boundedness*. Manuscript on arXiv: 1812.07500, 2018.

4.6.4 *Governed decompositions*

Bonamy and Pilipczuk prove in 2020 that graphs of bounded diversity are <u>polynomially</u> χ - bounded. For this purpose they derive the lemma below.

To present this lemma we need some definitions.

Definition 4.75. A generalized decomposition of a graph G is a pair (T, η) where T is a rooted tree and η a map $\eta : V(G) \to V(T)$.

Let (T, η) be a generalized decomposition. For an edge $e = \{u, v\} \in E(G)$ let $\eta(e)$ be the least common ancestor of u and v in T. For nodes $x, y \in V(T)$ write $x \preceq y$ when x is an ancestor of y. For a node $x \in V(T)$ define the graph $G <x>$ as the graph with the following sets of vertices and edges.

$$\begin{aligned} V <x> &= \{ y \in V(G) \mid x \preceq \eta(y) \} \\ E <x> &= \{ e \in E(G) \mid \eta(e) = x \} \end{aligned}$$

Definition 4.76. Let $\mathcal{C}$ be a class of graphs. A generalized decomposition (T, η) of a graph is <u>$\mathcal{C}$ - governed</u> if the graph $G <x> \in \mathcal{C}$ for all $x \in V(T)$.

Let (T, η) be a generalized decomposition of a graph. The (generalized) <u>diversity</u> of a branch rooted at a node $x \in V(T)$ is the number of neighborhood - classes of $G <x>$ — that is — the number of equivalence classes when two vertices of $G <x>$ are equivalent if they have the same neighbors in $V(G) \setminus V <x>$.

Let (T, η) be a generalized decomposition of diversity k. A <u>tagging</u> is a set of functions $\lambda_x : V <x> \rightarrow [k]$ $(x \in V(T))$ such that u and v in $V <x>$ have the same neighbors in $V(G) \setminus V <x>$ when $\lambda_x(u) = \lambda_x(v)$.

Let $e = \{x, y\} \in E(T)$ and let x be the parent of y. Let $\rho(e) : [k] \rightarrow [k]$ be a function which assigns $\rho(e)(i) = j$ if

$$u \in V <y> \quad \text{and} \quad \lambda_y(u) = i \quad \Rightarrow \quad \lambda_x(u) = j.$$

> The functions $\rho(e)$ $(e \in E(T))$ form a labeling of $E(T)$ with elements of $\mathcal{F} = [k]^{[k]}$.

Let $\mathcal{F} = [k]^{[k]}$ — that is — $\mathcal{F}$ is the collection of functions $[k] \rightarrow [k]$. Denote the composition of two functions f and g in $\mathcal{F}$ as $f \circ g$.

Definition 4.77. A subset $A \subseteq \mathcal{F}$ is forward Ramsey if for all $e, f \in A$

$$e = e \circ f.$$

Definition 4.78. A decomposition (T, η) of a graph is <u>Kruskalian</u> if the set of edge - labels is forward Ramsey.

A decomposition (T, η) is <u>shallow</u> if every path to the root has at most two nodes.

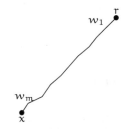

Figure 4.6: Edge - labels are elements of $\mathcal{F} = [k]^{[k]}$. A word w_x is the sequence of edge labels from the root to x. It relates to the function $w_1 \circ \cdots \circ w_m$.

Bonamy and Pilipczuk call the Kruskalian decompositions: splendid.

Bonamy and Pilipczuk prove the following lemma. We take a look at the proof.

Lemma 4.79. *Let $\mathcal{C}$ be a hereditary class of graphs. There exists $p \in 2^{O(k \log k)}$ and a sequence of hereditary classes*

$$\mathcal{D}_0 \quad \subseteq \quad \mathcal{D}_1 \quad \cdots \quad \subseteq \quad \mathcal{D}_p$$

such that

1. $\mathcal{D}_0 = \mathcal{C}$

2. $\mathcal{D}_p$ *is the class of graphs that have a $\mathcal{C}$ - governed decomposition of generalized diversity* k

3. *for* $i \in [p]$ *all graphs in* $\mathcal{D}_i$ *have a* $\mathcal{D}_{i-1}$ *- governed decomposition of generalized diversity* k *which is Kruskalian or shallow.*

4.6.5 Forward Ramsey splits

T. Colcombet, A combinatorial theorem for trees. In Proceedings ICALP'07, Springer, LNCS 4596, pp. 901–912, 2007.

To prove Lemma 4.79 the authors make use of a lemma that they attribute to Thomas Colcombet. We present the lemma without proof.

Consider a tree T with a root and an edge - labeling $\rho : E(T) \to \mathcal{F}$ where $\mathcal{F} = [k]^{[k]}$. A 'word' is a sequence of edge labels on a path from the root to a node.

Let $\mathcal{F}^*$ the set of words of finite length with letters in $\mathcal{F}$. For a nonempty word $w \in \mathcal{F}^*$ let $\phi(w) \in \mathcal{F}$ denote the function that is the composition of the letters in w.

Let w be a nonempty word say $w = w_1 \cdots w_n$. Let $w[x, y]$ denote the word $w_{x+1} \cdots w_y$ (for $0 \leqslant x < y \leqslant n$). A <u>split</u> of height h of w is a map $\{0, \cdots, n\} \to [h]$. Two positions $0 \leqslant x < y \leqslant n$ are s - equivalent if

$$s(x) \quad = \quad s(y) \quad \text{and} \quad s(z) \quad \leqslant \quad s(x) \qquad \text{for all } x \leqslant z \leqslant y.$$

Definition 4.80. A split is forward Ramsey if for every two s -
equivalent pairs $x < y$ and $x' < y'$

$$\phi(w[x,y]) \quad = \quad \phi(w[x,y]) \quad \circ \quad \phi(w[x',y']).$$

Lemma 4.81 (Colcombet). *There exists a map $\mu : \mathcal{F}^* \to [k^k]$ with
the following property. Let w be a nonempty word and let s_w be the
split $s_w(x) = \mu(w[0,x])$. Then s_w is forward Ramsey.*

For the tree T define a split of height h as a function $V(T) \to$
$[h]$. For a node x let w_x be sequence of edge - labels on the path
from the root to x. The split is forward Ramsey if it is a forward
Ramsey split of w_x for every $x \in V(T)$.

4.6.6 *Factorization of trees*

Let (T, r) be a rooted tree. A factorization is a partition of $V(T)$ in
parts that induce subtrees. The parts are called factors. The root
of a factor is the vertex closest to r.

Let $\mathcal{P}$ be a factorization. The quotient tree $T/\mathcal{P}$ is obtained from
T by shrinking each factor to its root.

Assume that T has an edge - labeling $\rho : E(T) \to \mathcal{F}$. The quotient
tree has an edge - labeling $\rho/\mathcal{P}$ defined as follows. Let $e = \{x, y\}$ be
an edge of $T/\mathcal{P}$ and assume that x is the parent of y. Let $e_1 \ldots e_m$
be the sequence of edges on the path in T from the root of x to the
root of y. Define the edge label of the quotient tree as follows

$$\rho/\mathcal{P}(e) \quad = \quad \rho(e_1) \circ \cdots \circ \rho(e_m).$$

Bonamy and Pilipczuk prove the following lemma.

Lemma 4.82. *There exists a sequence $(\mathcal{T}_i)_{i=0}^{3|\mathcal{F}|}$ which satisfies the
following.* [80]

- *each $\mathcal{T}_i$ is a class of edge - labeled trees with labels from $\mathcal{F}$*

- *for all i $\mathcal{T}_i \subseteq \mathcal{T}_{i+1}$*

Notice that we define the
'forward Ramsey - concept'
for

1. Sets $A \subseteq \mathcal{F}$

2. Splits of words in $\mathcal{F}^*$

3. Splits of $\mathcal{F}$ - labeled trees.

[80] The sequence has $3 \cdot k^k + 1$
elements.

- *the class $\mathcal{T}_0$ has only one element which is a tree with one node. The class $\mathcal{T}_{3|\mathcal{F}|}$ is the class of all edge - labeled trees with labels from $\mathcal{F}$*

- *for all $i > 0$ every tree in $\mathcal{T}_i$ has a factorization with factors in $\mathcal{T}_{i-1}$ and with a quotient tree that is either Kruskalian or shallow.*

Proof. Let (T, ρ) be a rooted tree with an $\mathcal{F}$ - labeling on its edges. Define the <u>level</u> of (T, ρ) as follows.

$$\text{level}\,(T, \rho) \quad = \quad \min \quad \{\, h \quad | \quad (T, \rho) \quad \text{has a split of height } h \,\}$$

The level of (T, ρ) is at most k^k by Colcombet's lemma.

For $x \in V(T)$ define $t(x) = \mu(w[0, x])$.

ALSO — define a <u>complexity</u> of (T, ρ) as follows.

- if T has only one node then then the complexity of (T, ρ) is 0

- otherwise — when T has at least two nodes — the complexity is the smallest number k such that (T, ρ) has a factorization of which every factor has complexity $< k$ and for which the quotient tree is Kruskalian or shallow.

We <u>claim</u> that

$$\text{complexity}\,(T, \rho) \quad \leqslant \quad 3 \cdot \text{level}\,(T, \rho). \qquad (\dagger)$$

THIS CLAIM IMPLIES THE LEMMA: let $\mathcal{T}_i$ be the set of trees with edge - labels in $\mathcal{F}$ that have complexity i.

We prove ($\dagger$) by induction on the level.

The following observation is our primary tool.

Exercise 4.59

Let $A, B \subseteq \mathcal{F}$ and assume that $A \cap B \neq \varnothing$. If A and B are forward Ramsey then so is $A \cup B$.

Recall the definition 4.77 of forward Ramsey - subsets of $\mathcal{F}$.

<u>Assume that the level of (T, ρ) is 1</u> — that is — T has a split of height one: $t(x) = 1$ for all $x \in V(T)$.

Let's start with the base case.

Exercise 4.60

Let $y \in V(T)$ be a child of the root and let $A_y \subseteq \mathcal{F}$ be those elements of $\mathcal{F}$ that are assigned by ρ to edges of the subtree rooted at y. The set A_y is forward Ramsey.

HINT: Let z be a child of y. The elements of w_z are forward Ramsey. All words w_z contain the element $\rho(r, y)$. The previous exercise shows that $A_y \cup \{\rho(r, y)\}$ is forward Ramsey. This implies that A_y is forward Ramsey.

Exercise 4.61

For every child y of the root the complexity of the tree rooted at y is at most one.

HINT: Take a factorization into single nodes. All factors have complexity zero and the quotient tree is just the tree T_y. By the previous exercise it is Kruskalian. This shows that the complexity of (T, ρ) is at most two because we can take a factorization with the root as one factor and each subtree rooted at a child y of r as a factor. Then the quotient is shallow.

This proves (†) for the base case; when the level of T is one.

Induction step: Assume that the level of (T, ρ) is $\ell > 1$ and let t be a split of T of height ℓ. Define $X \subseteq V(T)$ as follows

$$X \quad = \quad \{ x \in V(T) \mid t(x) \quad = \quad \ell \}.$$

Define a factorization $\mathcal{P}$ of T as follows. Two nodes x and y are in the same part if either

1. neither x nor y has an ancestor in X (which implies $x, y \notin X$)

2. x and y have the same least ancestor in X.

For $x \in X$ let $\mathcal{P}_x$ be the factorization of the subtree rooted at x.

Let $y \in X$ have an ancestor in $X \setminus y$. Then $(T_y/\mathcal{P}_y, \rho_y/\mathcal{P}_y)$ is Kruskalian. The argument to prove this is similar to the one asked for in Exercise 4.60: Let $x \in X$ be the least ancestor of y not equal to y and let $z \in X$ be a descendant of y in X. Let Q_{xz} be the path in $T/\mathcal{P}$ from x to z; then $\{x, y\}$ is the first edge of Q_{xz}. The set B_z

That base case was easy enough; let's get on with the induction step; I bet it goes 'in the same way.' (After all; we only have a short list of ingredients to cook the proof.)

Figure 4.7: An artist's impression of a factorization.

of elements assigned by $\rho/\mathcal{P}$ to edges of Q_{xz} is forward Ramsey. It follows that the union of $\cup B_z$ for $z \in X$ that are descendants of y is forward Ramsey.

CLAIM: Let $y \in X$ and assume that y has an ancestor in $X \setminus y$. Then the complexity of (T_y, ρ_y) is at most $3\ell - 1$. By the previous observation $(T_y/\mathcal{P}_y, \rho_y/\mathcal{P}_y)$ is Kruskalian. Let F be a factor of $\mathcal{P}_y$. We show that F has complexity at most $3\ell - 2$. The only node of F that is at height ℓ is the root of F. Take a factorization with the root as one factor and with every branch of the root as one factor. Each branch has level $\leqslant \ell - 1$ — and so — (by the induction hypothesis) its complexity is at most $3\ell - 3$. The root is a factor with one node and so it has complexity zero. The factorization is shallow — and so — F has complexity at most $3\ell - 1$.

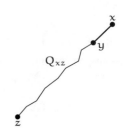

Figure 4.8: Illustration to show that $(T_y/\mathcal{P}_y, \rho_y/\mathcal{P}_y)$ is Kruskalian.

Let R be the subtree of T induced by the nodes that have at most one ancestor in X.

R Is the maximal subtree of T in which no path to the root has two elements of X.

CLAIM: The complexity of R is at most $3 \cdot \ell - 1$.

To see that notice that for every $x \in R \cap X$ the subtree R_x has complexity at most $3\ell - 2$ (because the only node of R_x of height ℓ is the root). [81]

[81] So — if the root of T is in X then we are done.

Let R' be the subtree of nodes that have no ancestor in X. Then R' has a forward Ramsey split of height $\ell-1$ and so its complexity is at most $3\ell - 3$.

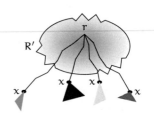

This proves the claim because $\{R', R_x \mid x \in R \cap X\}$ is a shallow factorization of R with all factors having complexity at most $3\ell - 2$.

Figure 4.9: The figure illustrates the factorization of R into R' and subtrees R_x.

IT'S TIME TO FINISH THE PROOF OF (†). Let $\mathcal{Q}$ be the partition of $V(T)$ with factors R and T_y for the nodes $y \in X$ that have exactly one ancestor $\neq y$ in X. Then all factors have complexity $\leqslant 3 \cdot \ell - 1$. — Furthermore — the quotient is shallow — and so — the complexity of T is at most $3 \cdot \ell$.   □

The lemma of Bonamy and Pilipczuk follows from Lemma 4.82.

Lemma 4.83. *Let $\mathcal{C}$ be a hereditary class of graphs and let $p = 3 \cdot k^k$. There exists a sequence of hereditary classes*

$$\mathcal{D}_0 \subseteq \mathcal{D}_1 \cdots \subseteq \mathcal{D}_p$$

such that

1. $\mathcal{D}_0 - \mathcal{C}$

2. $\mathcal{D}_p$ *is the class of graphs that have a $\mathcal{C}$ - governed decomposition of generalized diversity k*

3. *for $i \in [p]$ all graphs in $\mathcal{D}_i$ have a $\mathcal{D}_{i-1}$ - governed decomposition of generalized diversity k which is Kruskalian or shallow.*

Proof. By Lemma 4.82 there exists a sequence $(\mathcal{T}_i)_1^{3|\mathcal{F}|}$ of trees with edge - labels in $\mathcal{F}$ such that trees in $\mathcal{T}_i$ have a factorization that has all factors in $\mathcal{T}_{i-1}$ and a quotient which is Kruskalian or shallow.

Let $\mathcal{D}_i$ be the class of graphs that have a $\mathcal{C}$ - governed decomposition of diversity k which is a tree in $\mathcal{T}_i$.

WE LEAVE IT TO THE READER to check that the classes $\mathcal{D}_i$ satisfy the required properties. □

EXERCISE: Show that the class of circle graphs is closed under vertex - minors.

Remark 4.84. Geelen et al shows that for any circle graph H the graphs that do not have H as a vertex - minor have bounded rankwidth. — So — these graphs are polynomially χ - bounded.

4.6.7 Kruskalian decompositions

WE END THIS CHAPTER with some remarks about the Kruskalian decompositions. These observations show that a class of graphs of bounded rankwidth is polynomially χ - bounded.

Let $\mathcal{C}$ be a hereditary class of graphs and denote the closure of $\mathcal{C}$ under modular substitution by $\mathcal{C}^*$. — For example— when $\mathcal{C}$

be the class of graphs that have at most two vertices then $\mathcal{C}^*$ is the class of cographs.

Notice that $\mathcal{C}^*$ is the class of graphs that have a $\mathcal{C}$ - governed decomposition of diversity one.

When $\mathcal{C}$ is polynomially χ - bounded then so is $\mathcal{C}^*$.

The fact that T is Kruskalian implies that there is a partition of $V(\mathsf{G})$ into at most k parts such that for all parts the decomposition has an edge - labeling which a <u>constant</u> function.

For each part G^i partition the levels of T in odd and even. This defines two spanning subgraphs of G^i — say G_0^i and G_1^i where G_0^i has the edges of G^i defined by the nodes in even levels of T and G_1^i has the edges of G^i defined by nodes in odd levels of T.

Since the labeling is a constant function G_0^i and G_1^i are both in $\mathcal{C}^*$.

4.6.8   Exercise

Exercise 4.62

Show that the class of AT - free graphs is linearly χ - bounded.

HINT: Every connected AT - free graph has a dominating pair — that is a pair $\{a, b\}$ of vertices with the property that every path that runs between a and b is a dominating set in the graph.

Permutation graphs and interval graphs are AT - free. These classes are perfect; so $\chi = \omega$. These classes are complements of comparability graphs. That is ; they are intersection graphs of continuous functions $f : [0, 1] \to \mathbb{R}$. (This implies that cocomparability graphs are AT - free.) AT - free graphs are not perfect as C_5 is AT - free.

Remark 4.85. Graphs of bounded <u>linear</u> rankwidth are linearly χ - bounded.

J. Nešetřil, P. Ossona de Mendez, R. Rabinovich and S. Siebertz, Linear rankwidth meets stability. Manuscript on arXiv: 1911.07748, 2019.

4.7   Clustered coloring

Hadwiger conjectures that graphs without K_t - minor can be colored with $t - 1$ colors. This has been proved for $t \leqslant 6$ (and it is open for $t \geqslant 7$).

The conjecture dates back to 1943.
EXERCISE: What are the graphs that do not have K_3 as a minor? Are they colorable with two colors? K_4 poses similar questions that you should be able to answer. The case $t = 5$ is equivalent to the 4-color theorem. (That is so because by Wagner's theorem every 4 - connected graph is planar if and only if it has no K_5 - minor.)

Definition 4.86. Let G be a graph. A <u>cluster coloring</u> of G with k colors and cluster number c is a map $f : (V) \to [k]$ such that for each $i \in [k]$ the graph induced by the vertices of color i has no component with more than c vertices.

Definition 4.87. Let $\mathcal{G}$ be a class of graphs. The <u>cluster chromatic number</u> of $\mathcal{G}$ is the least number k for which there exists $c \in \mathbb{N}$ such that all graphs in $\mathcal{G}$ can be cluster colored with k colors and cluster number c.

In 2018 Jan van den Heuvel and David Wood proved the theorem below. In this chapter we take a look at the proof.

Theorem 4.88. *Let* $t \geqslant 4$*. Every* K_t *minor free - graph has a cluster coloring that uses* $(2t - 2)$ *colors and has cluster number* $\lceil (t-2)/2 \rceil$*.*

4.7.1   Bandwidth and BFS - trees with few leaves

To prove Theorem 4.88 we start with some easy exercises. (The rest turns out to be easy as well.)

Definition 4.89. A graph has bandwidth k if its vertices can be ordered $v_1 \cdots v_n$ such that

$$\{v_i, v_j\} \in E \quad \Rightarrow \quad |i - j| \leqslant k.$$

CLEARLY when a graph has bandwidth k then it is a subgraph of an interval graph with clique number $k + 1$ — that is — the pathwidth of a graph is at most its bandwidth.

Let G be a graph and let T be a BFS - tree of G. The tree partitions the vertices of G in layers [82] — say $V_0 \cdots V_\ell$ — where V_i is the set of vertices that are in G at distance i from the root. For a vertex $x \in V_i$ its <u>parent</u> is its neighbor in T in layer V_{i-1}.

The BFS - tree <u>orders each layer</u> V_i such that

1. the parent of a vertex $x \in V_i$ is the first neighbor of x in V_{i-1}

2. if $\{x, y\} \in E(G)$ and $x \in V_i$ and $y \in V_{i-1}$ then there does not exist $\{a, b\} \in E$ with a before x in V_i and b after y in V_{i-1}.

Exercise 4.63

Let $k \in \mathbb{N}$ and let G be a connected graph that has a BFS - tree with at most k leaves. Show that the bandwidth of G is at most k.

HINT: Notice that each layer has at most k vertices. Take a linear order

$$V_0 \quad V_1 \quad \cdots \quad V_\ell.$$

Let G be a graph and let T be a BFS - tree of G. A <u>BFS - subtree</u> of T is a subtree of T that contains the root of T. (A BFS - subtree is a BFS - tree of the graph induced in G by its vertices.)

Exercise 4.64

Let G be a connected graph and let T be a BFS - tree of G. Let S be a BFS - subtree of T with k leaves. Every vertex of G has at most 2k neighbors in $V(S)$.

HINT: Let P be a path in S that runs from the root to a leaf. It is sufficient to show that every vertex x in G has at most 2 neighbors in P. That is clearly so when $x \in P$ because P is a shortest path. Let $x \notin P$ and assume that x is adjacent to three vertices in P — say — a, b and c. Let $a \in V_{i-1}$, $b \in V_i$ and $c \in V_{i+1}$. Let y be the parent of x. Then — by the first rule — y appears before a and (by the second rule) x appears before b in V_i. But then (by the first rule) c is adjacent to x instead of b in T.

4.7.2 Connected partitions

In their paper Van den Heuvel and Wood prove the following theorem which implies Theorem 4.88.

Let G be a graph. A <u>connected partition</u> $\{H_1, \cdots, H_\ell\}$ is an ordered partition of $V(G)$ such that each part H_i induces a connected subgraph of G. Two parts are connected if there is an edge in G with an endpoint in both.

Theorem 4.90. *Let* $t \geqslant 4$ *and let* G *be* K_t - *minor free. Then* G *has a connected partition* $\{H_1, \cdots, H_\ell\}$ *which satisfies for all* i:

- H_i *is adjacent to at most* $t - 2$ *parts of* $H_1 \cdots H_{i-1}$

- *every vertex of the induced subgraph* H_i *has degree at most* $t-2$

- *the induced subgraph* H_i *is 2 - colorable with cluster* $\lceil (t-2)/2 \rceil$.

Exercise 4.65

Show that Theorem 4.90 implies Theorem 4.88.

HINT: Color the parts H_i such that adjacent parts receive different colors. — Obviously — $t-1$ colors are sufficient. Color the vertices of the graph G with a <u>pair</u> of colors: one element of the pair is the color of the part that contains the vertex and the other element of the pair is the color that the vertex receives by a 2 - coloring of the part with cluster number $\lceil (t-2)/2 \rceil$.

THEOREM 4.90 IS PROVED *via* the following two lemmas.

Lemma 4.91. *Let* G *be a connected graph and let* $A \subseteq V(G)$ *with* $|A| = k \geqslant 2$. *Let* H *be a connected induced subgraph of* G *with* $A \subseteq V(H)$ *and*

In an early paper Van den Heuvel et al. show that each part H_i has a BFS - tree with at most $t - 3$ leaves. — So — a 2 - coloring of the BFS - layers yields a 2 - coloring of the part with cluster number $t - 3$.

The Steiner tree problem asks for a tree in G (of minimum weight) which spans all vertices of a set of 'terminals.' This problem is fixed - parameter tractable (with parameter the number of terminals).

A 'minimal' induced connected subgraph H as specified in Lemma 4.91 is computable by a greedy algorithm.

- *every vertex of* $H - A$ *is a cutvertex of* H

- *for every vertex* $x \in H - A$ *every component of* $H - x$ *has a vertex of* A.

Then the graph H *has the following properties.*

1. *every tree contained in* H *has at most* k *leaves*

2. *every vertex of* H *has degree at most* k

3. H *has bandwidth at most* $k - 1$

4. H *has a 2 - coloring with cluster* $\lceil k/2 \rceil$

5. H *can be colored with colors red and blue such that*

   - *the red subgraph of* H *has at most* $k - 2$ *vertices*

   - *the blue subgraph of* H *is a union of at most* $k - 1$ *paths.*

Proof. All the leaves of a spanning tree T of H are in A (by the minimality of H). It follows that any tree in H has at most k leaves — and so — every vertex of H has degree at most k. By Exercise 4.63 H has bandwidth $k - 1$ (since H has a BFS - tree with a root in A).

We show that H has a 2 - coloring with cluster $\lceil k/2 \rceil$. When $|V(H)| = k$ then this is obvious. We proceed by induction on $|V(H)|$.

When $|V(H)| > k$ then H has a cutvertex. Let $v \notin A$ be a cutvertex such that some component L of $H - v$ has the least number of elements in A. — Clearly — $V(L) \subseteq A$ and $|V(L)| \leqslant \frac{k}{2}$.

Every vertex of $H - A$ is a cutvertex of H.

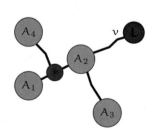

Let $H' = H - L$ and let

$$A' = (A \setminus V(L)) \cup \{v\}.$$

Then H' is a minimal connected subgraph of G that spans A' and $|V(H')| < |V(H)|$ and $|A'| \leqslant k$. By induction H' has a 2 - coloring with cluster number $\lceil k/2 \rceil$. To color H color the vertices of L by the color that is not used by v.

The figure illustrates the induced subgraph H.

We show that H has the required red / blue - coloring.

We use induction on k. When $k = 2$ then H is a path between two vertices of A. In that case color all vertices of H blue.

Assume $k \geqslant 3$ and let $x \in A$ be such that $H - x$ is connected. Let H' be a minimal connected subgraph of $H - x$ that contains $A \setminus x$. By induction H' can be colored with $\leqslant k - 3$ red vertices and $\leqslant k - 2$ blue paths. When $x \in V(H')$ then we are done.

Otherwise when $x \notin V(H')$ let $P = [x \cdots u, v, w]$ be a shortest path $x \rightsquigarrow A \setminus x$ in H — that is — v is the only vertex of $P \setminus w$ that has neighbors in $V(H')$. Color v red and the path $P \setminus \{v, w\}$ blue. This colors all vertices of H with a color red or blue. There are at most $k - 2$ red vertices and the blue vertices form a union of at most $k - 1$ paths. — Furthermore — $V(H) \subseteq V(P) \cup V(H')$ (by the minimality of H).

This proves the lemma.   □

Lemma 4.92. *Let G be a connected graph, let $A \subseteq V(G)$ and let $|A| = k \geqslant 2$. There exists an induced subgraph H which satisfies all the items mentioned in Lemma 4.91 and which — furthermore — has the property that every vertex in G has at most $2k - 2$ neighbors in $V(H)$.*

To compute H as mentioned in Lemma 4.92 construct a BFS - tree rooted at a vertex of A. Extract a minimal subtree that spans all vertices of A. Repeatedly remove vertices that are not in A and that are not cutvertices.

Proof. Let T be a BFS - tree with a root in A. Let S be the minimal subtree of T that contains all elements of A. Let H be a minimal induced subgraph of $G[V(S)]$ — as mentioned in Lemma 4.91 — which spans A. Then $V(H) \subseteq V(S)$.

The tree S has at most $k - 1$ leaves. By Exercise 4.64 every vertex of G has at most $2(k - 1)$ neighbors in $V(S)$.

The claims follow from Lemma 4.91.

This proves the lemma.   □

4.7.3   A decomposition of K_t minor free graphs

IN THIS SECTION WE PROVE THEOREM 4.90 (which implies
Theorem 4.88).

Definition 4.93. Let $\{H_1, \cdots, H_\ell\}$ be a connected partition of
a graph G. It has <u>width</u> k if for all t every component of
$G - \bigcup_{t=1}^{i} H_t$ is adjacent to at most k of the graphs $H_1, \cdots, H_i$.

The following theorem implies Theorem 4.90.

Theorem 4.94. *Let* $t \geqslant 4$ *and let* G *be a graph without* K_t *as a
minor. There is a connected partition* $\{H_1, \cdots, H_\ell\}$ *of width* $t - 2$
such that each H_i *satisfies the following conditions.*

(a) H_i *is a graph of which every vertex has degree at most* $t - 2$

(b) H_i *has bandwidth* $t - 3$

(c) H_i *has a 2 - coloring with cluster* $\lceil (t-2)/2 \rceil$

(d) H_i *can be colored with colors red and blue such that*

- *there are at most* $t - 4$ *red vertices*
- *the blue vertices induce a union of at most* $t - 3$ *paths.*

— *Furthermore* — *let* C *be a component of* $G - \bigcup_{t=1}^{i} H_t$. *Then*
C *satisfies the following.*

(i) *at most* $t - 2$ *of the graphs* $H_1, \cdots, H_i$ *are adjacent to* C *and
those are pairwise adjacent*

(ii) *every vertex of* C *has at most* $2t - 6$ *neighbors in each of the
sets* $V(H_1), \cdots, V(H_i)$.

Proof. We may assume that G is connected. Construct the H_i
one by one as follows. Choose an arbitrary vertex x of G and let
$V(H_1) = \{x\}$. Then H_1 and every component of $G - H_1$ satisfy all
the items.

Assume there is a component C in $G - \bigcup_{t=1}^{i} H_t$. Let $Q_1, \cdots, Q_k$
be the elements of $\{H_1, \cdots, H_i\}$ that are adjacent to C. We may
assume that $k \leqslant t - 2$ and that the Q_i's are pairwise adjacent.

Define H_{i+1} as follows. For $j \in [k]$ let $v_j \in C$ be a vertex with a neighbor in Q_j. When $k = 1$ then let $V(H_{i+1}) = \{v_1\}$. When $k \geqslant 2$ let H_{i+1} be the graph as produced in Lemma 4.92 which contains $\{v_1, \cdots, v_k\}$.

Let C' be a component of $G - \bigcup_{t=1}^{i+1} H_t$. Notice that either $C' \subset C$ or $C \cap C' = \varnothing$.

Assume $C' \cap C = \varnothing$. Then C' is a component of $G - \bigcup_{t=1}^{i} H_t$ and C' is not adjacent to H_{i+1}.

Assume $C' \subset C$. By induction and Lemma 4.92 every vertex of C' has at most $2t - 6$ neighbors in each of $H_1, \cdots, H_{i+1}$. The neighbors of C' are a subset of $Q_1, \cdots, Q_k, H_{i+1}$ and these are pairwise adjacent.

Suppose $k = t - 2$. Contract each of $Q_1, \cdots, Q_{t-2}, H_{i+1}$ and C' to a single vertex. This produces K_t which contradicts that G does not have K_t as a minor.

This proves the theorem.   □

4.7.4 Further reading

IN CASE YOU HAVEN'T READ THIS; it's a "golden oldie."

S. Dreyfus and R. Wagner, The Steiner problem in graphs, *Networks* **1** (1972), pp. 195–207.

In 2018 Chung - Hung Liu and Sang-il Oum found that the cluster chromatic number of K_t - minor graph graphs is at most $3(t-1)$. [83]

C.-H. Liu and S.-i. Oum, Partitioning H - minor free graphs into three subgraphs with no large components, *Journal of Combinatorial Theory* **128** (2018), pp. 114–133.

In 2020 Chun - Hung Liu determines the cluster chromatic number up to a small additive constant for graphs without H - immersion.

[83] The cluster numbers of these colorings are very large.

Chun - Hung Liu, Immersions and clustered coloring. Manuscript on arXiv: 2007.00259, 2020.

Definition 4.95. A coloring of a graph with k colors and defect d is a coloring $f : V(G) \to [k]$ such that in every monochromatic component every vertex has degree at most d.

In their paper Van den Heuvel and Wood show that K_t - minor free graphs can be colored with $t-1$ colors and defect $t-2$. Edwards et al. show that the class of K_t - minor free graphs has defective chromatic number equal to $t-1$.

A class of graphs has defective chromatic number k if there exists $d \in \mathbb{N}$ such that every graph in the class has a coloring with k colors and defect d.

The paper of Edwards et al. shows — also — that the class of graphs without topological K_t minor is colorable with $t-1$ colors and defect $O(t^4)$.

K. Edwards, D. Y. Kang, J. Kim, S.-i. Oum and P. Seymour, A relative of Hadwiger's conjecture, *SIAM Journal on Discrete Mathematics* **29** (2015), pp. 2385 – 2388.

LET T BE A TREE WITH n EDGES. When n is large enough K_{2n+1} can be packed with $2n+1$ copies of T.

The edges of K_{2n+1} can be colored with $2n+1$ colors such that each color induces a copy of T. This is called Ringel's conjecture.

R. Montgomery, A. Prokovskiy and B. Sudakov, A proof of Ringel's conjecture. Manuscript on arXiv:2001.02665, 2020.

Let T be an oriented tree on n vertices. When n is large enough any tournament with $2n-1$ vertices contains a copy of T.

D. Kühn, R. Mycroft and D. Osthus, A proof of Sumner's universal tournament conjecture for large tournaments. Manuscript on arXiv: 1010.4430, 2010.

4.8   Well - Quasi Orders

We might as well put the definitions of well - quasi orders here.

Definition 4.96. A quasi - order is a set with binary relation which is reflexive and transitive.

Notice that a quasi - order is similar to a partial order — except that — quasi - orders are not necessarily anti - symmetric.

Definition 4.97. A quasi - order is a well - quasi order if — for any infinite sequence of elements $x_1, x_2, \cdots$ — there exist indices $i < j$ such that $x_i \preceq x_j$.

Exercise 4.66

Let T be a tree not necessarily finite with a root. Define the run - out of a point x in T as the supremum of the lengths of paths in T that have x as endpoint.

Prove Kőnig's infinity lemma:
If every point of T has finite out - degree and some point P has infinite run - out then there is an infinite path that starts in P. Notice that the condition that the out - degree is finite is essential.

Hint: At any point P with infinite run - out there must be a successor Q which has also infinite run - out.

This proof is not constructive; only a proof by contradiction shows the existence of a successor. So it is not a proof in the sense of L.E.J. Brouwer.

4.8.1   Higman's Lemma

LET A BE A FINITE ALPHABET OF LETTERS and consider an infinite sequence
$$w_1 \quad w_2 \quad w_3 \quad \cdots$$
of 'words' — that is — finite nonempty sequences of letters
$$w_i = w_i[1] \quad w_i[2] \quad \ldots \quad w_i[k]$$

where k is the length of the word w_i and $w_i(\ell) \in A$ for $\ell \in [k]$. [84]
Higman's lemma asserts that there exist $i < j$ such that w_i is a
subsequence of w_j. By that we mean that there is an increasing
function $f : \mathbb{N} \to \mathbb{N}$ such that

$$w_i[k] = w_j[f(k)]$$

for $k = 1$ up to the length of w_i.

Below we present Nash-Williams' proof of the lemma.

Lemma 4.98. *The set* A^* *of finite nonempty sequences over a
finite alphabet* A *is well-quasi ordered by the subsequence relation.*

Proof. Nash-Williams introduces the notion of a 'bad sequence.'
A sequence (w_i) $(i \in \mathbb{N})$ is bad if for no pair $i < j$ w_i is a
subsequence of w_j. Assume that there exists a bad sequence.

Construct a <u>minimal</u> bad sequence as follows. Let $x_1 \in A^*$ be
a word of minimal length that starts some bad sequence. Let
$x_2 \in A^*$ be a word of minimal length such that there exists a bad
sequence that starts with x_1, x_2. For $i \in \mathbb{N}$ let $x_{i+1} \in A^*$ be of
minimal length such that there is a bad sequence starting with
$x_1 \cdots x_{i+1}$.

Choose an infinite subsequence of (x_i) — say (y_i) — such that all
words y_i start with the same letter. From each y_i remove the first
letter and call the new sequence (y_i'). Then (y_i') is a bad sequence.

Let $y_1 = x_n$. The sequence

$$x_1 \quad \cdots \quad x_{n-1} \quad y_1' \quad y_2' \quad \cdots$$

is bad also. But this contradicts the choice of x_n since y_1' is a
shorter word and $x_1 \cdots x_{n-1} y_1'$ starts a bad sequence.

This proves the lemma.   □

[84] The notation A^* is used
for the set of finite sequences
over the alphabet A. So we
have $w_i \in A^*$ for $i \in \mathbb{N}$.
Perhaps we should have used
'k_i' instead of k for the
length of the word w_i. Just
note that words may have different lengths. We assume
that the words are finite and
nonempty so their lengths
are in $\mathbb{N}$.

4.8.2   Kruskal's Theorem

Kruskal's theorem extends Higman's lemma to sequences of labeled trees.

Consider an infinite sequence of rooted trees [85] — say

$$\mathsf{T}_1 \quad \mathsf{T}_2 \quad \cdots$$

Assume that the vertices of each tree have been labeled with elements from some finite set — say $[\mathsf{k}]$ — for $\mathsf{k} \in \mathbb{N}$. [86]

Kruskal's theorem says that there exist $\mathsf{i} < \mathsf{j}$ such that T_i can be 'embedded' into T_j. This is defined as follows.

Definition 4.99. Define for two labeled trees T_i and T_j

$$\mathsf{T}_\mathsf{i} \quad \preceq \quad \mathsf{T}_\mathsf{j}$$

if there is an injective map

$$\mathsf{f} : \mathsf{V}(\mathsf{T}_\mathsf{i}) \quad \rightarrow \quad \mathsf{V}(\mathsf{T}_\mathsf{j})$$

that satisfies

1. f maps the root of T_i to the root of T_j

2. f preserves labels — that is —

$$\mathsf{label}(\mathsf{f}(\mathsf{x})) = \mathsf{label}(\mathsf{x}),$$

3. for any two vertices x and y of T_i their common ancestor is mapped to the common ancestor of $\mathsf{f}(\mathsf{x})$ and $\mathsf{f}(\mathsf{y})$.

We present Kruskal's theorem without proof.

Theorem 4.100. *The set of rooted trees with vertices labeled from a finite set is well - quasi ordered. — That is — let* $\mathsf{k} \in \mathbb{N}$ *and let* $\mathsf{T}_1, \mathsf{T}_2, \cdots$ *be an infinite sequence of rooted trees with vertices labeled from* $[\mathsf{k}]$. *Then there exist indices* $\mathsf{i} < \mathsf{j}$ *that satisfy*

$$\mathsf{T}_\mathsf{i} \preceq \mathsf{T}_\mathsf{j}.$$

The theorem remains true when the trees are labeled with elements of a well - quasi order. Nash–Williams gives an elegant proof of Kruskal's theorem.

[85] A root in a tree is one vertex that is labeled as 'root.'

[86] To spell: 'labeling' and 'labelling' are both OK.

A function $\mathsf{A} \rightarrow \mathsf{B}$ is _injective_ if every $\mathsf{b} \in \mathsf{B}$ is the image of at most one element in A.

4.8.3   Gap embeddings

Let $(Q, \ll)$ be a quasi - order and let $k \in \mathbb{N}$. Let $\mathcal{T}$ be the set of triples (T, f, a) where

- T is a rooted tree

- $f : V(T) \to Q$

- $a : E(T) \to [k]$.

Define a quasi - order $\preceq$ on $\mathcal{T}$ as follows. Let $t = (T, f, a)$ and $r = (R, g, b)$ be two elements of $\mathcal{T}$. Let $t \preceq r$ if there exists an injective map $\eta : V(T) \to V(R)$ which satisfies the following conditions.

1. η maps the root of T to the root of R and η maps the common ancestor of any two points in T to the common ancestor of their images in R

2. for an $x \in V(T)$: $g(\eta(x)) \ll f(x)$

3. for any edge $e = \{x, y\} \in E(T)$: $a(e) \leqslant b(e')$ for all $e' \in E(R)$ that lie on the path from $\eta(x)$ to $\eta(y)$.

Theorem 4.101. *Let $(Q, \leqslant)$ be a well - quasi order and let $k \in \mathbb{N}$. The collection $\mathcal{T}$ of labeled trees — as defined above — is well - quasi -ordered by the relation $\preceq$.*

The theorem remains valid when $\mathbb{N}$ is replaced with — say — $\mathbb{N} \cup \{i, 0, \infty\}$.

Remark 4.102. Above we assume that the edge - labels are from a totally ordered set $[k]$. Tzameret shows that this can be relaxed — but — <u>not</u> 'all the way.'

I. Tzameret, *Kruskal - Friedman gap embedding theorems over well - quasi - orderings*. Thesis, Tel Aviv University 2002.

4.9 Threshold graphs and threshold - width

THRESHOLD GRAPHS WERE GIVEN THEIR NAME BY CHVÁTAL AND HAMMER. Below we present one way to define this class of graphs.

Definition 4.103. A graph is a <u>threshold graph</u> if every induced subgraph has an isolated vertex or a universal vertex. [87]

[87] A vertex is isolated if its neighborhood is empty. A vertex is universal if it is adjacent to all other vertices. When a graph has only one vertex it is both; isolated and universal.

Exercise 4.67

A graph is a threshold graph if and only if its complement is that. Show that a graph is a threshold graph if and only if it has no induced P_4, C_4, or $2K_2$.

Exercise 4.68

Figure 4.10: The figure shows P_4, C_4 and $2K_2$.

A graph is a <u>split graph</u> if its vertices partition into a clique and an independent set. Show that a graph is a split graph if and only if it does not contain $2K_2$, C_4 or C_5 as an induced subgraph. Show that every threshold graph is a split graph.

The <u>threshold dimension</u> $\theta(G)$ of a graph $G = (V, E)$ is the minimum $k \in \mathbb{N}$ for which there are k threshold graphs $G_i = (V, E_i)$ with $\bigcup E_i = E$. There exists an $O(n^3)$ - algorithm to check if $\theta(G) \leqslant 2$. To check if $\theta(G) \leqslant 3$ is NP-complete.

Exercise 4.69

What is the threshold dimension of a trivially perfect graph? [88]

[88] Recall from Section 2.9.3 on page 80 that a graph is trivially perfect if it has no induced P_4 and no induced C_4. Since $\bar{C}_4 = 2K_2$ a graph is a threshold graph if and only if it and its complement are trivially perfect.

4.9.1   Threshold - width

A — SOMEWHAT — SIMILAR CONCEPT is that of the threshold - width of a graph.

Definition 4.104. A graph G has <u>threshold - width</u> $\leqslant k$ if it has k independent sets $N_1, \cdots, N_k$ such that there is a threshold graph H that has G as a spanning subgraph and every edge of H — which is not an edge of G — has both endpoints in N_i for some $i \in [k]$.

Denote the smallest k such that a graph G has threshold - width $\leqslant k$ as $\tau(G)$. In this section we show that $\tau \leqslant k$ is fixed parameter - tractable. [89]

[89] EVEN BETTER!

Theorem 4.105. *There exists a characterization of the graphs that satisfy $\tau \leqslant k$ by a finite collection of forbidden induced subgraphs.*

Proof. Observe that the class of graphs that satisfy $\tau \leqslant k$ is hereditary. — So — there exists a collection $\mathcal{F}$ of graphs F that satisfy $\tau(F) > k$ and for every vertex $x \in V(F)$ $\tau(F - x) \leqslant k$.

IT IS OUR JOB TO PROVE THAT $|\mathcal{F}|$ IS FINITE.

A graph is a threshold - graph if its vertices can be put in a linear order such that each vertex is either adjacent to all vertices that come after it or not adjacent to any vertex that comes after it.

Let G be a graph that has threshold - width $\leqslant k$. Let $N_1, \cdots, N_k$ be k independent sets in G that witness this. We identify the graph with a <u>word</u> with letters from a finite alphabet as follows.

Label each vertex x with a $(0, 1)$-vector $\ell(x)$ of length k with the ith entry 1 if $x \in N_i$. Then G has a linear order of its vertices such that each vertex x is — either

(i) not adjacent to any vertex that comes after it

(ii) adjacent to exactly all vertices y that come after x for which
$\ell(x) \cdot \ell(y) = 0$.

(Here $\ell(x) \cdot \ell(y)$ denotes the inner product of the two labels; it is not
zero when x and y occupy a similar independent set N_i.)

Identify a graph with a sequence of labels in an elimination order
as above. Notice that if a graph H has a sequence which is a
subsequence of the sequence of the graph G then H is isomorphic
to an induced subgraph of G. [90]

We proceed by contradiction. Let

$$F_1 \quad F_2 \quad \cdots \tag{4.14}$$

be an infinite sequence of different elements of $\mathcal{F}$. Each graph F
in this sequence has $\tau(F) > k$ — but — we can choose an arbitrary
vertex r in it such that $\tau(F - r) \leqslant k$.

Denote a choice for a vertex in the graph F_i as r_i. Each $F_i - r_i$
identifies with a word via the labeling procedure described above
with letters from a finite alphabet.

Extend the labels of the vertices of $F_i - r_i$ with one additional
$(0, 1)$-label: 1 if the vertex is adjacent to r_i and 0 otherwise. Notice
that if a graph $F_i - r_i$ has a sequence which is a subsequence of a
graph $F_j - r_j$ then F_i is isomorphic to an induced subgraph of F_j.
[91]

Replace the sequence (4.14) by a sequence of words of finite length
with letters from a finite alphabet. We can make use of Higman's
Lemma (Lemma 4.98 on Page 214): the sequence must contain
elements $i < j$ such that the word F_i is a subsequence of the word F_j
— that is — F_i is isomorphic to an induced subgraph of F_j. This is
a contradiction — so — $|\mathcal{F}| \in \mathbb{N}$.

This proves the theorem. $\qquad \square$

Lemma 4.106. *For any graph its rankwidth is at most 2^τ.*

Proof. Let G be a graph, let $N_1, \cdots, N_k$ be k independent sets
in G, and let H be an embedding of G in a threshold graph with
every $e \in E(H) \setminus E(G)$ contained in some N_i.

The graph H has rankwidth 1 (it is distance - hereditary). Let $\mathcal{C}$ be a carving of H with $\mathsf{rank}(X) \leqslant 1$ for every $X \in \mathcal{C}$. The rank of the cut matrix $(X, V \setminus X)$ of the adjacency matrix of G is at most 2^k. To see that observe that each independent set N_i is a 0-submatrix in the adjacency matrix of G. It follows that for $X \in \mathcal{C}$ there are at most 2^k different neighborhoods in $V \setminus X$. [92]

[92] This bound is sharp.

This proves the lemma. $\square$

Corollary 4.107. *Problems that can be formulated in monadic second-order logic can be solved in $O(n^3)$ time for graphs of bounded threshold - width.*

Theorem 4.108. *Threshold -width is fixed - parameter tractable.*

Proof. The class of graphs with threshold - width $\leqslant k$ is characterized by a finite collection of forbidden induced subgraphs. This shows that the recognition can be formulated in monadic second-order logic.

The graphs of threshold - width $\leqslant k$ have rankwidth at most 2^k. By Courcelle's theorem there exists an $O(n^3)$ - algorithm to recognize graphs of threshold - width $\leqslant k$.

This proves the theorem. $\square$

Remark 4.109. Theorem 4.108 can also be proved via the formulation of an elimination order — that is — a graph has threshold - width $\leqslant k$ if and only if every induced subgraph has an isolated vertex or a vertex x which is adjacent to all other vertices except those that are in one of the k independent sets that contain x. This property be formulated in monadic second order logic.

4.9.2   On the complexity of threshold - width

In this section we show that computing the threshold - width of a graph is NP-complete.

LET US LOOK AT AN 'EASIER' PROBLEM — FIRST. Let $\mathcal{K}$ be the class of all cliques ie all complete graphs. The $\underline{\mathcal{K} \text{ - width}}$ of a graph is the minimum number of independent sets $N_1, \cdots, N_k$ in the graph such that every nonedge of G has its endpoints in one of the N_i. [93]

Lemma 4.110. $\mathcal{K}$ - Width is NP-complete.

Proof. The problem is equivalent to finding a cover of the edges of $\bar{G}$ with a minimal number of cliques. Kou, Stockmeyer, and Wong proved that this is NP-complete. [94]   □

Theorem 4.111. *Threshold - width is NP-complete.*

Proof. We reduce $\mathcal{K}$-width to threshold - width.

Let G be a graph for which we would like to compute the $\mathcal{K}$-width. Starting with G construct the graph G':

(a) add a clique C with n^2 vertices and make every vertex of C adjacent to every vertex of G

(b) add one more vertex ω and make it adjacent to all vertices of G.

CLEARLY if we add an edge between every nonadjacent pair in G then G' becomes a threshold graph. We claim that this is the best way to embed G' into a threshold graph.

Let x and y be a nonadjacent pair in G. When x and y are not adjacent in a threshold embedding of G' then ω is adjacent to all vertices of C in that embedding. That is so because a threshold graph has no C_4. HOWEVER to make ω adjacent to all vertices of C needs n^2 independent sets (since C is a clique).

[93] We could call this the 'clique - width' of the graph. Unfortunately, there is another concept closely related to rankwidth that carries that name. So we just call it $\mathcal{K}$ - width.

[94] L. Kou, L. Stockmeyer and C. Wong, *Covering edges by cliques with regard to keyword conflicts and intersection graphs.* Communications of the ACM **21** (1978), pp. 135–139.

This proves that the threshold - width of G' equals the $\mathcal{K}$ - width of G.

By Lemma 4.110 this proves the theorem.   □

4.9.3 *A fixed - parameter algorithm for threshold - width*

We have shown that threshold - width is fixed - parameter tractable.

IN THIS SECTION WE PRESENT AN ALGORITHM.

To get in the mood we start with some — easy — exercises.

Exercise 4.70

Show that a graph is a threshold graph if and only if for any pair of its vertices x and y

$$N(x) \subseteq N[y] \quad \text{or} \quad N(y) \subseteq N[x].$$

IN OTHER WORDS a graph is a threshold graph if and only if its vertices can be put into a linear order

$$x_1 \quad \cdots \quad x_n$$

such that

$$1 \leqslant i < j \leqslant n \quad \Rightarrow \quad N(x_i) \subseteq N[x_j].$$

Exercise 4.71

Let G be a graph and let $\{N_1, \cdots, N_k\}$ be a 'witness' — ie — a collection of k independent sets in G. Design an algorithm to check if G can be embedded into a threshold graph H such that every edge of H which is not an edge of G has both endpoints in some N_i. HINT: Define a k-universal vertex as a vertex for which the sets N_i that contain it cover all its nonneighbors.

Exercise 4.72

Let G be a graph of threshold - width $\leqslant k$. Let $N_1, \cdots, N_k$ be k independent sets in G that witness this. The witness $\{N_1, \cdots, N_k\}$ is <u>well-linked</u> if every N_i is a maximal independent set in G. Prove that every graph of threshold - width $\leqslant k$ has a well-linked witness.

Exercise 4.73

Assume G has a well-linked witness $\{N_1, \cdots, N_k\}$ and a threshold embedding H. Label each vertex as in the proof of Theorem 4.105 with a vector $\ell(x)$ of length k. Assume two vertices x and y satisfy $N_H(x) \subseteq N_H[y]$. Then

$$N_G(x) \subseteq N_G[y] \quad \Leftrightarrow \quad \ell(x) \geqslant \ell(y).$$

Definition 4.112. Let $k \in \mathbb{N}$. A set M of vertices in a graph is called a <u>k-probe module</u> if either

1. $|M| \geqslant 3$ and every pair of vertices in M is a false twin (in the graph)

2. $|M| \geqslant k + 3$ and every pair of vertices in M is a true twin.

Lemma 4.113. *Let* G *be a graph; let* $k \in \mathbb{N}$; *and let* M *be a* k-*probe module in* G. *Then for any* $x \in M$

$$\tau(G) \leqslant k \quad \Leftrightarrow \quad \tau(G - x) \leqslant k.$$

Proof. — CLEARLY — $\tau(G - x) \leqslant \tau(G)$ for any $x \in V(G)$. Assume $\tau(G - x) \leqslant k$ for some x in a k-probe module M. Let H be a threshold embedding of $G - x$. Since H is a split graph its vertices partition into a clique — say C — and an independent set.

First assume that M is an independent set - module. Then

$$|M \setminus x| \quad \geqslant \quad 2.$$

Let $y \in M \backslash x$. If y is in the independent set (that is, if $y \in V(H) \backslash C$) then we can let x be a false twin of y. This produces a threshold embedding of G of width $\leqslant k$. When $(M \backslash x) \subseteq C$ then let x be a true twin of an arbitrary element of $M \backslash x$. — Again — this produces a threshold embedding of G with width $\leqslant k$.

Assume that M is a clique module. Then it has at least $k+3$ elements. At least $k+1$ of those are in C. Choose $z \in M \cap C$ such that $N_H[z]$ is minimal. Assume that z has a neighbor u in H which is not a neighbor of z in G. Then u is a neighbor in H but not in G of every vertex of $M \cap C$. Since M is a clique in G the vertex u is contained in at least $k+1$ independent sets. This is a contradiction.

— So —

$$N_H[z] \quad = \quad N_G[z]$$

and we can let x be a twin of z.

This proves the lemma.   □

Definition 4.114. A vertex x is <u>maximal</u> if for all $y \in V$

$$N(y) \subseteq N(x) \quad \Rightarrow \quad N(y) = N(x) \quad \text{and}$$
$$N[y] \subseteq N[x] \quad \Rightarrow \quad N[y] = N[x].$$

Lemma 4.115. *Let G be a graph with threshold - width $\leqslant k$ and assume that G has no k-probe module. Then the number of maximal vertices in G is at most*

$$2^{k+1} + k.$$

Proof. Let H be a threshold embedding of G with a well-linked witness $\{N_1, \cdots, N_k\}$.

Partition the vertices of H into equivalence classes $M_0, M_1, \cdots$ of vertices with the same open - or closed neighborhood. (Each M_i is a clique or an independent set in H.) Order the classes such that for each $x_i \in M_i$ and $x_{i+1} \in M_{i+1}$:

$$N_H(x_{i+1}) \subseteq N_H[x_i].$$

(So when H is connected M_0 is its set of universal vertices.)

Partition each M_i into sets of vertices that have the same label.[95] These 'label-sets' are modules in G. Since there is no k-probe module each label-set has at most 2 vertices when it is independent and at most $k + 2$ when it is a clique. It follows that for each i

$$|M_i| \leqslant 2(2^k - 1) + (k + 2) = 2^{k+1} + k.$$

Notice that there are at most 2^k label-sets of maximal vertices. At most $2^k - 1$ are in independent sets N_i and they have at most 2 vertices. At most one is a clique and it has at most $k + 2$ vertices.

This proves the lemma. □

Lemma 4.116. *Let G be a graph of threshold - width $\leqslant k$. Assume that G has no isolated vertices and no k-probe modules. There exists a set $Y \subseteq V$, $|Y| \leqslant 2^{2(k+1)}$, such that every threshold embedding that is a witness has its set of universal vertices $M_0 \subseteq Y$. The set Y can be computed in linear time.*

Proof. Since G has no isolated vertices H is connected. Let M_0 be its set of universal vertices.

To compute a set Y that contains M_0 start with $Y = \varnothing$. Repeatedly add the set of vertices to Y that are maximal in G and remove those from the graph.

After at most 2^k repetitions each label-set of M_0 is contained in Y. Each set of maximal elements has size at most $2^{k+1} + k$ which shows

$$|Y| \leqslant 2^k \cdot (2^{k+1} + k) \leqslant 2^{2(k+1)}$$

This proves the lemma. □

Exercise 4.74

Let U be a set of labeled vertices. As usual; each label is a $(0,1)$-vector of length k and for each entry i the vertices — say N_i — that have a 1 in that entry form an independent set.

The set U is a <u>probe clique</u> if the inner product of any two labels of elements in U is zero — that is — if the nonedges of U are exactly the pairs that share some N_i.

Call $U \subseteq V(G)$ <u>probe universal</u> if each $x \notin U$ can be given a label such that $U \cup x$ is a probe clique.

Let U be a probe universal set and let $x \notin U$ be such that the set

$$U' \quad = \quad U \cup N(x)$$

can be labeled as probe universal with the same number of nonempty label-sets as U. When there is such a vertex then choose x such that $N(x)$ is minimal. When G has an embedding with U as a universal set then G has an embedding with U' as a universal set.

Theorem 4.117. *Let $k \in \mathbb{N}$. There exists an $O(n^2)$ algorithm that recognizes graphs of threshold width $\leqslant k$.*

Proof. We may assume that G has no isolated vertices or k-probe modules.

By Lemma 4.116 there is a constant number of feasible universal sets.

Assume there exists a vertex x that can be labeled such that $N(x)$ extends the universal set in such a way that it does not increase the number of label-sets. By Exercise 4.74 the algorithm can safely extend the probe universal set with $N(x)$. Next the algorithm removes the vertex x and tries to find another greedy extension.

When there are no more greedy extension the algorithm computes the set Y as in Lemma 4.116. It then tries all subsets of Y as possible extensions of the probe universal set.

There can be at most 2^k steps in this algorithm that increase the number of label-sets in the probe universal set. The set Y of maximal elements can be computed in $O(n^2)$ time.

This proves the theorem.   □

Exercise 4.75

Define a width - parameter for the class of distance-hereditary graphs as follows. A graph G has DH-width $\leqslant k$ if it has k independent sets $N_1, \cdots, N_k$ and an embedding H which is distance-hereditary such that every edge of H which is not an edge of G has both endpoints in some N_i.

Is DH-width fixed-parameter tractable?

HINT: Is there a monadic second-order formulation of $DH(G) \leqslant k$?

4.10   Black and white - coloring

CLAUDE BERGE POSED THE PROBLEM to put b black queens and w white queens on a chess board so that no two queens of opposite colors hit each other. [96]

Definition 4.118. Let G be a graph and let $b, w \in \mathbb{N}$. A <u>black and white - coloring</u> of G chooses b black vertices and w white vertices such that no black vertex is adjacent to any white vertex.

[96] We talk about Western chess played on a 8×8 board. A black and white queen hit each other if they are placed in the same row, column, or diagonal provided no other piece is placed between them.

Exercise 4.76

Show that the black and white - coloring problem can be solved in linear time on graphs of bounded treewidth.

Exercise 4.77

Show that there is a polynomial-time algorithm to solve the black and white - coloring problem on cographs.

Hint: Define a boolean variable $\phi(G, b, w)$ with value true if the graph G has a black and white coloring with b black vertices and w white vertices.

Let G_1 and G_2 are two cographs and let G be the join or union of the two. Express $\phi(G, w, b)$ as a function of $\phi(G_1, w_1, b_1)$ and $\phi(G_2, w_2, b_2)$.

This leads to an (n^5) algorithm to check if a cograph has a black and white coloring (for any values b and w). Improve your algorithm to solve the black and white - coloring problem on cographs so that it runs in $O(n^3)$.

4.10.1   The complexity of black and white - coloring

In this section we show that the black and white - coloring problem is NP-complete even when restricted to the class of splitgraphs (see Exercise 4.68 for the definition of a splitgraph.) [97]

Theorem 4.119. *The black and white coloring - problem is* NP-*complete on splitgraphs.*

Proof. Splitgraphs are closed under complementations. The 'inverse black and white coloring - problem' is similar except that it is required that every black vertex is adjacent to every white vertex.

Let G be a graph. Construct a splitgraph $H = (S \cup C, E')$ as follows. The clique C of H is $V(G)$. The independent set S of H is $E(G)$. A vertex of C is adjacent to a vertex of S in H if the vertex is not an endpoint of the edge in G.

Let Ω be a maximum clique in G of size k and let

$$V' = V(G) \setminus \Omega.$$

The computation of the clique number in G remains NP-complete when n is even, $k = \frac{n}{2}$ and $n > 6$. [98] — Henceforth — we

[97] A graph is a splitgraph if it is a clique or an independent set or else V partitions in $\{C, S\}$ where C induces a clique and S a stable set. A graph is a splitgraph if it has no induced C_4, C_5 or $2K_2$.

[98] D. Johnson, *The NP-completeness column — an ongoing guide*, Journal of Algorithms **8** (1987), pp. 438–448.

assume that.

Notice that H has an inverse black and white coloring with

$$b = k \quad \text{and} \quad w = k + \binom{k}{2} = \binom{k+1}{2} \qquad (4.15)$$

To see that — color all vertices of V' black. They are in H adjacent to all edges in S that have both ends in Ω and to all vertices of $C \setminus V'$.

For the converse assume that the splitgraph H has an inverse black and white coloring with b and w as in Equation (4.15). Since S is an independent set all colored vertices of S are the same color.

First assume that S contains no white vertices. Then C contains a set W of white vertices and vertices in $C \setminus W$ are black. However $w > 2k = n$ (since $n > 6$) — so — this is not possible. So the inverse black and white - coloring has white vertices in S.

Let S_w be the set of white vertices in S and let $V' \subseteq C$ be the set of k black vertices. If an edge of G is in S_w then it is not incident with a vertex of V'. All those edges are incident with vertices in $V(G) \setminus V'$. Since $|V \setminus V'| = k$ and $|S_w| = \binom{k}{2}$ the only possibility is that S_w is the set of edges of a k-clique $V \setminus V'$.

This proves the theorem. □

4.11 k – Cographs

In this section we illustrate the importance of Kruskal's theorem.

Recall the definition of a cograph; Definition 2.78. [99]
By Theorem 2.79 (on Page 84) cographs can be encoded into cotrees.

Definition 4.120. Let G be a graph. A <u>cotree</u> for G is a pair (T, f) — where

1. T is a rooted binary tree [100]

2. f : $V(G) \to$ leaves(T) is a bijection that identifies each vertex of G with one leaf of T

[99] A graph is a cograph if it has no induced P_4.

Figure 4.11: P_4

[100] A rooted tree is <u>binary</u> if either it has only one point (which is then both root and leaf) or the root has degree 2, all leaves have degree 1, and all other vertices have degree 3. Notice that there is no binary tree with two points.

3. each internal node of T is labeled as $\otimes$ or $\oplus$

4.

$$\{x, y\} \in E(G) \quad \Leftrightarrow$$

the least common ancestor of $f(x)$ and $f(y)$ in T is labeled $\otimes$.

$$(4.16)$$

By Theorem 2.79 a graph is a cograph if and only if it has a cotree.

We <u>parametrize</u> the class of cographs as follows. [101]

Definition 4.121. Let $k \in \mathbb{N}$. A graph G is a <u>k-cograph</u> if it has a decomposition (T, f) such that

I. T is a rooted binary tree   ,

II. each leaf of T is labeled with an element from $[k]$

III. f is a bijection $V(G) \to \mathsf{leaves}(\mathsf{T})$

IV. all internal nodes are labeled by a a symmetric binary relation on $[k]$ (eg represented by a symmetric Boolean $k \times k$-matrix)

V.

$$\{x, y\} \in E(G) \quad \Leftrightarrow$$

the lowest common ancestor of $f(x)$ and $f(y)$ is labeled σ,

with σ such that

$$\sigma(\mathsf{label}(f(x)), \mathsf{label}(f(y))) = \text{true}. \quad (4.17)$$

— Notice that — ordinary cographs are 1-cographs.

[101] To spell, both 'parametrize' and 'parameterize' are OK.

Exercise 4.78

Show that — for each $k \in \mathbb{N}$ — the class of k-cographs is hereditary.

Hint: Let G be a k-cograph and let (T, f) be a decomposition tree for G. For an induced subgraph H consider the subtree of T that contains all vertices of H.

Exercise 4.79

Show that a class of k–cographs is closed under creating twins. A <u>twin</u> is a pair of vertices — say x and y — such that [102]

$$N[x] = N[y] \quad \text{if } x \text{ and } y \text{ are adjacent}$$
$$N(x) = N(y) \quad \text{otherwise.}$$

[102] When the pair is adjacent, the twin is called a true twin. When the pair is non-adjacent, it is called a false twin.

When G is a k–cograph — and H is obtained from G by creating a twin of some vertex in G — then H is also a k–cograph.

4.11.1   Recognition of k – Cographs

In this section we show that there exists a characterization of k–cographs by a finite set of forbidden induced subgraphs.

For $k \in \mathbb{N}$ denote the class of k-cographs as $\mathcal{C}(k)$.

Theorem 4.122. *There exists a finite set of graphs S_k — such that — a graph $G \in \mathcal{C}(k)$ if and only if G contains no element of S_k as an induced subgraph.*

Proof. The set S_k is the set of inclusion–minimal graphs that are not k–cographs. We show that S_k is finite.

Assume that S_k is not finite. Then we can choose an infinite sequence of pairwise different graphs in S_k — say

$$G_1, G_2, \cdots .$$

Since each graph G_i is inclusion–minimal — we have that $G_i - r_i$ is a k–cograph — for each vertex $r_i \in V(G_i)$. Pick one arbitrary vertex r_i in each G_i.

Consider a sequence of rooted binary trees

$$T_1, T_2, \cdots$$

where T_i is a k–cotree for the graph $G_i - r_i$.

Extend the labels of the vertices in each T_i as follows. In T_i, give a vertex $z \in V(G_i - r_i)$ an extra label

$$\text{label}(z) = \begin{cases} 0 & \text{if } z \text{ is not adjacent to } r_i, \text{ and} \\ 1 & \text{otherwise}. \end{cases}$$

— By Kruskal's theorem — the newly labeled sequence of trees $T_1, T_2, \cdots$ satisfies

$$\exists_i \ \exists_j \ i, j \in \mathbb{N} \quad \text{and} \quad i < j \quad \text{and} \quad T_i \preceq T_j.$$

But — owing to the new labeling — this implies that G_i is an induced subgraph of G_j — which is a contradiction.

This proves the theorem.   □

Corollary 4.123. *For* $k \in \mathbb{N}$, *there exists a polynomial – time algorithm that recognizes* k *– cographs.* [103]

This follows from the fact that S_k is finite. Namely we can test whether one element of S_k —say with t vertices — is an induced subgraph of a graph in $O(n^{t+2})$ time. So the recognition takes time

$$O\left(|S_k| \cdot n^{t+2}\right) = O\left(n^{t+2}\right),$$

$$\text{where } t = \max\{|V(S)| \mid S \in S_k\}. \quad (4.18)$$

[103] But ... nobody knows what it is, as long as S_k is unknown.

4.11.2 Recognition of k – Cographs — revisited

Actually we can do much better. If I gave you the definition, it would be easy for you to check that k–cographs have <u>rankwidth</u> at most k. Courcelle showed that

every problem that can be formulated in MS_1 can be solved in $O(n^3)$ time for graphs of rankwidth at most k.

It is an easy exercise to show that — for any graph S — there is a monadic second–order formula expressing that S is an induced subgraph of a graph G.

Theorem 4.124. *The recognition of* k-*cographs is fixed parameter – tractable. There exists an* $O(n^3)$ *algorithm to tests if a graph is a* k-*cograph.*

Exercise 4.80

Prove that any graph is a k-cograph, for some $k \in \mathbb{N}$. Define the

$$\text{cograph}-\text{width}(G) = \min\{k \mid G \text{ is a } k-\text{cograph}\}.$$

Show that $\text{cograph}-\text{width}$ is fixed parameter – tractable — that is — show that there exists an $O(f(k) \cdot n^3)$ algorithm that checks if the $\text{cograph}-\text{width}$ is at most k (for some function $f: \mathbb{N} \to \mathbb{N}$.

4.11.3 Treewidth of Cographs

Exercise 4.81

Prove the following lemma.

Lemma 4.125. *Let a graph* G *be the join of two graphs — say*

$$G = G_1 \otimes G_2.$$

Then

$$\text{tw}(G) = \min\{\text{tw}(G_1) + |V(G_2)|, \text{tw}(G_2) + |V(G_1)|\}.$$

Hint: Let H be a chordal embedding of G. If both H_1 and H_2 have nonadjacent vertices H has a C_4.

Exercise 4.82

Prove the following lemma.

Lemma 4.126. *Let a graph* G *be the union of two graphs — say that* $G = G_1 \oplus G_2$. *Then*

$$\text{tw}(G) = \max\{\text{tw}(G_1), \text{tw}(G_2)\}.$$

Exercise 4.83

Design a linear–time algorithm that computes the treewidth of cographs. You may assume that a cotree is part of the input.

4.12   Minors

IT IS A BIT UNFORTUNATE that graphs are not well-quasi-ordered by the induced subgraph relation. — For example — Figure 4.12 shows an infinite sequence of graphs, $T_1, \cdots$ that are pairwise <u>incomparable</u> [104] with respect to the induced subgraph relation — that is —

$$\forall_{i \neq j} \quad \neg (T_i \preceq_{\text{ind}} T_j),$$

where $\preceq_{\text{ind}}$ is the induced subgraph relation.   (4.19)

The way to avoid any infinite sequence of incomparable graphs, is to define a minor–ordering.

Definition 4.127. A graph H is a <u>minor</u> of a graph G if there is a sequence of

1. vertex deletions

2. edge deletions and

3. edge contractions

— performed on the graph G — that turns it into H.

[104] they are pairwise 'dissimilar.'

;-)

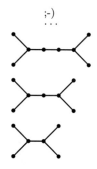

Figure 4.12: A sequence of graphs that is not well-quasi-ordered by the induced subgraph relation.

Exercise 4.84

Prove that a graph H is a minor of a graph G if and only if $V(G)$ can be partitioned into sets

$$V_1, V_2, \cdots, V_h \quad \text{where} \quad V(H) = [h]$$

— such that —

1. each $G[V_i]$ is connected

2.

$$\{i, j\} \in E(H) \quad \Rightarrow \quad \exists_{v_i \in V_i} \; \exists_{v_j \in V_j} \; \{v_i, v_j\} \in E(G).$$

Hint: — Since each $G[V_i]$ is connected — it can be contracted to one vertex. The second condition guarantees that H is a subgraph of the remainder. Notice that this proves the following corollary.

Corollary 4.128. *For any graph* H *the property that a graph* G *contains* H *as a minor can be formulated in* MS_1.

4.12.1 The Graph Minor Theorem

Robertson and Seymour proved the following theorem — which extends Kruskal's.

Theorem 4.129. *The class of all graphs is well-quasi-ordered by the minor relation.*

— Equivalently — we have the following result.

Theorem 4.130. *Every class of graphs that is closed under minors has a finite* obstruction set.

Let $\mathcal{G}$ be a class of graphs that is closed under taking minors. — So —

$$G \in \mathcal{G} \quad \text{and} \quad H \preceq_{minor} G \quad \Rightarrow \quad H \in \mathcal{G}.$$

Then there is a <u>finite</u> set $\mathcal{O}$ of graphs — called the obstruction set — which characterizes $\mathcal{G}$ in the following sense:

$$G \in \mathcal{G} \quad \Leftrightarrow \quad \forall_{O \in \mathcal{O}} \; \neg (O \preceq_{minor} G).$$

To see that Theorem 4.130 follows from Theorem 4.129, let $\mathcal{G}$ be a class of graphs that is closed under minors. Let $\mathcal{O}$ [105]

---

[105] $\mathcal{O}$ is the set of minimal elements, under the minor relation, that are not in $\mathcal{G}$.

be the set of graphs O that satisfy

$$O \notin \mathcal{G} \quad \text{and} \quad \forall_{O'} \quad (O' \preceq_{\text{minor}} O \quad \text{and} \quad O' \neq O) \quad \Rightarrow \quad O' \in \mathcal{G}.$$

If $|\mathcal{O}|$ is infinite [106] we can choose an infinite sequence $O_1, \cdots$ of graphs in $\mathcal{O}$ that are all different. — However — by Theorem 4.129 there must exist

$$i < j \quad \text{such that} \quad O_i \preceq_{\text{minor}} O_j$$

which is a contradiction.

This proves Theorem 4.130.

[106] ∞

Exercise 4.85

Show that the following classes of graphs are closed under taking minors;

1. the class of planar graphs [107]

2. the class of graphs with treewidth at most k, for $k \in \mathbb{N}$.

What is the obstruction set for the class of planar graphs?
Hint: Recall Kuratowski's theorem — A graph is planar if and only if it has no element of

$$\{ K_5, K_{3,3} \}$$

as a minor. [108]

[107] A graph is planar if it can be drawn in the plane without crossing edges. A plane graph is a planar graph together with an embedding of it in the plane.

[108] Harary dedicated his book to Kuratowski, "who gave K_5 and $K_{3,3}$ to those who thought planarity was nothing but topology."

4.13 General Partition Graphs

Definition 4.131. A graph G is a <u>general partition graph</u> if there exists a set S and a map which assigns a subset $S_x \subseteq S$ to every vertex $x \in V$ such that [109]

1. for all pairs of vertices x and y

$$\{x, y\} \in E \quad \Leftrightarrow \quad S_x \cap S_y \neq \varnothing$$

[109] The graph is a partition graph if it satisfies the three conditions and furthermore no two S_x and S_y ($x \neq y$) are the same.

2. $S = \cup_x S_x$

3. when M is a maximal independent set then

$$\{ S_m \mid m \in M \} \qquad \text{is a partition of } S.$$

> In this section we show that for every class of graphs which is not the class of all graphs and which is closed under taking minors there is a polynomial - time algorithm to check if a graph in the class is a general partition graph.

So — for example — there exists an efficient algorithm to check if a planar graph is a general partition graph.

Exercise 4.86

Show that a graph is a general partition graph if and only if it has a set $\mathcal{C}$ of cliques such that

(a) every edge $\{x, y\} \in E$ has both endpoints in some clique $C \in \mathcal{C}$ — that is — $\mathcal{C}$ covers the edges of G

(b) every maximal independent set hits every $C \in \mathcal{C}$.

Exercise 4.87

Show that every cograph is a general partition graph.

Exercise 4.88

Let G be a graph. Let the graph H be obtained from G by adding one vertex to every edge in G and making that adjacent to the two endpoints of the edge. Then H has only one clique - cover with $|E(G)|$ maximal cliques. Furthermore every maximal independent set hits every clique in this cover. So H is a general partition graph.

AT SOME POINT IN HISTORY it was discovered that general partition graphs satisfy the triangle condition: [110]

[110] It can be shown that an AT-free graph satisfies the triangle condition if and only if it is a general partition graph. This can be checked on AT-free graphs in polynomial time. It is conjectured that the triangle condition is co-NP-complete.

> A graph satisfies the triangle condition if for every maximal independent set M and every edge $\{x, y\}$ in $G - M$ there is a vertex $m \in M$ such that $\{x, y, m\}$ is a triangle in G.

Exercise 4.89

Design an algorithm to check if a planar graph satisfies the triangle condition. [111]

[111] T. Kloks, C. Lee, J. Liu and H. Müller, *On the recognition of general partition graphs*. Proceedings WG 2003, Springer-Verlag, Lecture Notes in Computer Science 2880 (2003), pp. 273–283.

Exercise 4.90

Not all graphs that satisfy the triangle condition are general partition graphs. For example the figure on Page 149 shows a circle graph that satisfies the triangle condition but is not a general partition graph.

Show that every general partition graph satisfies the triangle condition.

Our claim that we can test if a graph is general partition for minor - closed classes follows easily from the following lemma.

Lemma 4.132. *Let* $k \in \mathbb{N}$ *and let* $\mathcal{G}$ *be a class of graphs that satisfy* $\omega \leqslant k$. *There exists a polynomial - time algorithm to check if a graph in* $\mathcal{G}$ *is a general partition graph.*

Proof. We use Exercise 4.86.

Let $\mathcal{G}$ be a class of graphs with clique number $\leqslant k$. — Clearly — graphs in $\mathcal{G}$ have only $O(n^k)$ maximal cliques and we can compute a list of maximal cliques in polynomial time — eg — via the algorithm of Bron and Kerbosch.

Let C be a maximal clique for which there is a maximal independent set M such that

$$C \cap M = \varnothing.$$

Then C is not in a clique cover of which every element is hit by every maximal independent set. Call C <u>intolerable</u>.

We can recognize whether C is intolerable in polynomial time as follows. Let $C = \{x_1, \cdots, x_\ell\}$. For every choice of $y_i \in N(x_i)$ check if $\{y_1, \cdots, y_\ell\}$ is an independent set. If there exists a choice $Y = \{y_1, \cdots, y_\ell\}$ which is independent then Y is contained in a maximal independent set M with $C \cap M = \varnothing$.

Let $\mathcal{C}$ be the set of tolerable cliques (ie those that are not intolerable). We have that G is a general partition graph if and only if $\mathcal{C}$ covers all edges of G.

This proves the lemma. $\square$

Corollary 4.133. *Let $\mathcal{G}$ be a class of graphs which is not the class of all graphs and which is closed under taking minors. There exists a polynomial - time algorithm to check if a graph in $\mathcal{G}$ is a general partition graph.*

Proof. By the graph minor theorem the class $\mathcal{G}$ has a finite obstruction set $\mathcal{F}$. This set is not empty since $\mathcal{G}$ does not contain all graphs. Let

$$k \quad = \quad \min\{|V(F)| \mid F \in \mathcal{F}\}.$$

No graph in $\mathcal{G}$ can have a clique of size $> k$ (since it would have $F \in \mathcal{F}$ as a subgraph). $\square$

Exercise 4.91

The <u>red maximal independent set problem</u> is the following. GIVEN a graph G and a coloring of its vertices with colors red and blue. QUESTION: does G have a maximal independent set with only red vertices?

This problem is NP-complete even when restricted to planar graphs. Show that there is a polynomial - time algorithm to solve RED MAXIMAL INDEPENDENT SET for graphs with $\omega \leqslant k$. [112]

[112] T. Kloks, D. Kratsch, C. Lee and J. Liu, *Improved bottleneck domination algorithms.* Discrete Applied Mathematics **154** (2006), pp. 1578 – 1592.

4.14   Tournaments

An <u>orientation</u> of a graph G assigns to every edge $\{x, y\} \in E(G)$ an orientation — say — either xy or yx. [113]

> [113] We may also write $x \to y$ instead of xy.

Definition 4.134. A <u>tournament</u> is an orientation of a complete graph.

4.14.1   Tournament games

TWO PLAYERS PLAY A GAME. The board they use is a tournament. They both choose a point of the tournament. When they chose the same point the outcome of the game is a draw. Otherwise the player who chose the head of the arc formed by the two chosen points is the winner. [114]

> [114] It's like the "paper, scissors and stone game."

Let T be a tournament. A <u>probability distribution</u> is a function $w : V(T) \to [0, 1]$ which satisfies

$$\sum_{x \in V(T)} w(x) \;=\; 1.$$

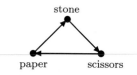

For a subset $S \subseteq V(T)$ we write $w(S) = \sum_{x \in S} w(x)$.

A player wishes to find a <u>winning</u> probability distribution — that is — he wishes to find a probability distribution which satisfies

Figure 4.13: The paper, scissors and stone game

$$\forall_{x \in V(T)} \quad w(I(x)) \;\geqslant\; w(O(x))$$

where $I(x)$ and $O(x)$ are the sets of vertices that beat x (so the arrows point from $I(x)$ towards x) and are beaten by x (the arrows point away from x towards $O(x)$). [115]

Fisher and Ryan show that every tournament has a winning probability distribution. In this section we show their proof.

> [115] A winning distribution has the property that for every $x \in V(T)$ it is at least as likely to beat x as it is to lose to x.

In their proof they make use of Farkas' lemma.

Lemma 4.135. *Given a matrix* M *and a vector* b *over the reals. Exactly one of the two following systems of linear inequalities has a solution.*

1. $Mx = b$ *and* $x \geqslant 0$

2. $M^T y \geqslant 0$ *and* $b^T y < 0$.

Let T be a tournament. The <u>payoff matrix</u> K has entries

$$k_{ij} \quad = \quad \begin{cases} 0 & \text{if } i = j \\ -1 & \text{if } i \rightarrow j \\ 1 & \text{if } i \leftarrow j. \end{cases}$$

Since T is a tournament $K^T = -K$ — that is — K is *skew-symmetric*.

Theorem 4.136. *Every tournament has a winning probability distribution. A winning distribution* w *satisfies*

$$w(x) > 0 \quad \Rightarrow \quad w(I(x)) \quad = \quad w(O(x)).$$

Proof. By definition; a distribution w is winning if

$$w \quad \geqslant \quad 0 \quad \text{and} \quad 1^T w \quad = \quad 1 \quad \text{and} \quad Kw \quad \leqslant \quad 0.$$

Assume that there is no winning distribution. Then the following system has no solution

$$\begin{pmatrix} K & I \\ 1^T & 0^T \end{pmatrix} \begin{pmatrix} w \\ z \end{pmatrix} = \begin{pmatrix} 0 \\ 1 \end{pmatrix} \quad \text{and} \quad \begin{pmatrix} w \\ z \end{pmatrix} \geqslant \begin{pmatrix} 0 \\ 0 \end{pmatrix}.$$

By Farkas' lemma the following system has a solution

$$\begin{pmatrix} -K & 1 \\ I & 0 \end{pmatrix} \begin{pmatrix} u \\ v \end{pmatrix} \geqslant \begin{pmatrix} 0 \\ 0 \end{pmatrix} \quad \text{and} \quad \begin{pmatrix} 0^T & 1 \end{pmatrix} \begin{pmatrix} u \\ v \end{pmatrix} < 0.$$

This implies $u \geqslant 0$ and $Ku < 0$. But then $w = u / (1^T u)$ is a winning probability distribution — a contradiction.

When w is winning $w_i (Kw)_i \leqslant 0$ for all i. Since K is skew-symmetric $w^T Kw = -w^T Kw = 0$ and this implies

$$w_i \quad > \quad 0 \quad \Rightarrow \quad w(I(i)) \quad = \quad w(O(i)).$$

This proves the theorem.   □

4.14.2 Trees in tournaments

Sumner conjectures that every tournament with $2(n-1)$ vertices contains any oriented tree with n vertices. This is true when n is large enough.

A weaker result was obtained by A. El Sahili in 2004.

Theorem 4.137 (El Sahili). *Every tournament with $3(n-1)$ vertices contains every oriented tree with n vertices.*

In this section we take a look at the proof.

MEDIAN ORDERS of digraphs were introduced by Havet and Thomassé. A median order of a digraph is defined as follows.

Definition 4.138. A <u>median order</u> of a digraph is an ordering of the vertices $v_1 \cdots v_n$ which maximizes

$$|\{\, (v_i, v_j) \in E \ \mid \ i < j \,\}|.$$

Exercise: Design an algorithm to compute a median order.

HINT: Notice that the problem is feedback arc set.

Exercise 4.92

Let $v_1 \cdots v_n$ be a median order of the vertices of a digraph.

1. any interval $v_{i+1} \cdots v_j$ is a median order of the digraph induced by $\{v_{i+1}, \cdots, v_j\}$

2. let $I = \{v_{i+1}, \cdots, v_j\}$. Then $|N^+(v_i) \cap I| \geqslant |N^-(v_i) \cap I|$.

Exercise: What are the median orders in graphs with a transitive orientation?

Definition 4.139. Let A and D be digraphs. An <u>embedding</u> of A in D is an <u>injection</u> $f : V(A) \to V(D)$ which satisfies

$$(a, b) \in E(A) \quad \Rightarrow \quad (f(a), f(b)) \in E(D).$$

A embeds in D if A is isomorphic to a subgraph of D.

Definition 4.140. Let A and D be digraphs and let $M = v_1 \cdots v_n$ be a median order of D. An embedding f of A in D is an M-embedding if for every <u>final segment</u> $I = \{v_{i+1} \cdots v_n\}$

$$|f(A) \cap I| \quad < \quad \frac{1}{2} \cdot |I| + 1.$$

Lemma 4.141. *Let* T *be a tournament with at least three vertices and let* $M = v_1 \cdots v_n$ *be a median order of* D. *Let* T' *be the tournament induced by* $\{v_1, \cdots, v_{n-2}\}$ *and let* $M' = v_1 \cdots v_{n-2}$. *Let* A *be a digraph with a leaf* (x, y). [116] *Assume that* $A - y$ *has an* M' *- embedding* f' *in* T'. *Then* A *has an* M *- embedding in* T *which extends* f'.

Proof. Let $A' = A - y$. Let f' be an M' - embedding of A' in T'. Let $f'(x) = v_i$ and let $I' = \{v_{i+1} \cdots v_{n-2}\}$. Then

$$|f'(A') \cap I'| \quad < \quad \frac{1}{2} \cdot |I'| + 1.$$

Let $I = \{v_{i+1} \cdots v_n\}$. Since T is a tournament and M is a median order

$$|N_T^+(v_i) \cap I| \quad \geqslant \quad \frac{1}{2} \cdot |I| = \frac{1}{2} \cdot |I'| + 1.$$

So

$$|N_T^+(v_i) \cap I| \quad > \quad |f'(A') \cap I'| \quad = \quad |f'(A') \cap I|.$$

We conclude that v_i has an outneighbor $v_j \in I \setminus f'(A')$ — say v.

Define $f(y) = v$ and $f = f'$ everywhere else. Then f is an M - embedding of A in T.

This proves the lemma. $\qquad \square$

A <u>branching</u> is a rooted tree with an orientation that is directed away from the root.

Exercise 4.93

Let A be a branching on n vertices and let T be a tournament with $2(n-1)$ vertices. Then A has an M - embedding in T for every median order M of T.

HINT: Use Lemma 4.141.

A. El Sahili remarks that this lemma is suggested in the paper by Havet and Thomassé on median orders.

[116] In general a leaf is a vertex with exactly one neighbor. Here we assume that the vertex y has exactly one in - neighbor x and no outneighbor.

When a digraph has a leaf y then it has a median order with y at the end.

Well - rooted trees

Definition 4.142. A digraph A is $\underline{t\text{-embeddable}}$ if A has an M - embedding in every tournament that has t vertices and median order M.

Definition 4.143. An oriented tree with a root is $\underline{well\text{-rooted}}$ if its root is a source.

Let A be a well - rooted tree. An edge of A is $\underline{backward}$ if its head is closer to the root. Let b be the number of backward edges.

The set of backward edges — if any — span a digraph — say B. Let c be the number of components of B.

FINALLY; let $d = b - c$.

Lemma 4.144. *Let A be a well - rooted tree with n vertices. Then A is $(2n + 2d)$ - embeddable with d defined as above.*

Proof. By induction on c. First assume $c = 0$. Then A is a branching and the claim follows from Exercise 4.93.

Assume $c > 0$. If A has a leaf y with parent x such that (x, y) is a forward arc then the claim follows from Lemma 4.141. So we may assume that $\underline{every}$ leaf of A is a source.

Let T be a tournament with $2(n + d)$ vertices and median order M — say $M = v_1 \cdots v_{2n+2d}$. We show that A has an M - embedding f in T.

Let B' be a component of B which contains a leaf of A. Let y be its root and let x be the parent of y. Then (x, y) is a forward arc.

Let $n' = |V(B')|$. Then $A - B'$ has $n - n'$ vertices and $b - (n' - 1)$ backward edges and $c - 1$ backward components.

Let T' be the tournament induced by the initial segment

$$M' \quad = \quad v_1 \quad \cdots \quad v_{2n+2d-4(n'-1)}.$$

If there are no backward edges the tree A is a branching.

B is a forest and it has has b edges. If c is the number of components, then what can you say about the number of vertices in B? (It's $b + c$.) B has no isolated vertices so $b > c$ unless both are zero.

Recall that A is a well - rooted tree — so — y is not the root.

By induction A' has an M' - embedding f' in T'.

Let A'' be obtained from A' by adding $2(n'_1)$ forward edges with tail x. Let S be the set of heads of these edges. By Lemma 4.141 A'' has an M - embedding f'' in T which extends f'.

Let U be the subtournament of T induced by $f''(S)$. Then U has $2(n'-1)$ vertices. By Exercise 4.93 U contains B'. Let g be an isomorphism from B' to a subtournament of U which is isomorphic to B'.

Define the map $f : V(A) \to V(T)$ as follows.

$$ f(x) \quad = \quad \begin{cases} f'(x) & \text{if } x \in V(A') \\ g(x) & \text{if } x \in B'. \end{cases} $$

Then f is an M - embedding of A in T.

This proves the lemma. $\square$

We now prove Theorem 4.137.

Theorem 4.145. *Every oriented tree with $n \geqslant 2$ vertices is $3(n-1)$ - embeddable.*

Proof. For any root that we choose in A we may assume that the number of forward arcs is at least the number of backward arcs — otherwise — we consider the problem of embedding the 'inverse' of A. [117]

We may assume that A is not a branching. Choose a root r in A which minimizes d. We may assume that r is a source. — To see that — assume there is a vertex v incident with an arc (v, r). If we choose v as the root then d decreases or else (if $\{v, r\}$ is one backward component) the tree is well rooted with v as a root and the same value d.

Since A is well - rooted we can apply Lemma 4.144: A is $2(n+d)$ - embeddable. We have

$$ b \;\leqslant\; \frac{n-1}{2} \quad \text{and} \quad c \;\geqslant\; 1 \quad \Rightarrow \quad d \;\leqslant\; \frac{n-3}{2}. $$

This proves the theorem. $\square$

[117] The inverse is obtained by replacing every arc (x, y) by its inverse (y, x).

Remark 4.146. In their paper Havet and Thomassé conjecture the following: Let A be an oriented tree with at most k leaves. Every tournament on $n + k - 1$ vertices contains A.

Further reading

F. Havet and S. Thomassé, Median orders of tournaments: a tool for the second neighborhood problem and Sumner's conjecture, *Journal of Graph Theory* **35** (2000), pp. 244–256.

D. Kühn, R. Mycroft and D. Osthus, An approximate version of Sumner's universal tournament conjecture. Manuscript on arXiv: 1010.4429, 2010.

4.14.3   Immersions in tournaments

Let G and H be digraphs. [118] The digraph H <u>immerses in</u> G if there is a map $\eta : H \to G$ which satisfies the following criteria.

[118] A digraph is an oriented graph. Each edge has an orientation; either xy or yx.

1. $\eta(x) \in V(G)$ for every $x \in V(H)$

2. when $x, y \in V(H)$ and $x \neq y$ then $\eta(x) \neq \eta(y)$

3. $\eta(xy)$ is a directed path in G from $\eta(x)$ to $\eta(y)$ for every edge $xy \in E(H)$

4. when e and f are edges of H and $e \neq f$ then $\eta(e)$ and $\eta(f)$ are edge - disjoint.

The digraph H <u>strongly</u> immerses in G when — additionally — the following condition is satisfied.

5. When $x \in V(H)$ and $e \in EH)$ and x is not an endpoint of e then $\eta(x)$ is not on the path $\eta(e)$.

To define immersions for graphs replace 'arc' with 'edge' in the definition.

Exercise 4.94

Let H and G be graphs. Show that H immerses in G if and only if H is an induced subgraph of a graph G' obtained from G *via* a sequence of edge lifts.

An edge - lift takes two edges that share an endpoint — say $\{x, a\}$ and $\{x, b\}$ — and replaces it with one edge $\{a, b\}$.

Chudnovsky and Seymour prove that tournaments are well - quasi ordered by strong immersion. In this section we review their proof.

The same is *not* true for digraphs. To see that consider the set of even length cycles and orient the edges so that there is no directed path with more than two vertices. No element of this set immerses in another one.

What happened earlier ..

In their paper Chudnovsky and Seymour use the following result (which they published in a separate paper). Let G be a digraph. A layout is a linear order of its vertices. Let

$$v_1 \quad \cdots \quad v_n$$

be a layout of G. The layout has cutwidth k if for each i there are at most k arcs that have their tail in $\{v_1, \cdots, v_i\}$ and their head in $\{v_{i+1}, \cdots, v_n\}$. The digraph G has cutwidth k if it has a layout of cutwidth k.

Theorem 4.147. *Let $\mathcal{S}$ be a set of tournaments. The following two statements are equivalent.*

1. *there exists $k \in \mathbb{N}$ such that all tournaments in $\mathcal{S}$ have cutwidth at most k*

2. *there exists a digraph H such that H does not strongly immerse in any tournament of $\mathcal{S}$.*

Remark 4.148. The two statements are also equivalent with this: there exists $k \in \mathbb{N}$ such that every vertex of a tournaments in S is in at most k edge - disjoint directed cycles.

Exercise 4.95

Let H be a cyclic triangle. The tournaments in which H does not immerse are the transitive tournaments. Transitive tournaments have cutwidth 0.

Figure 4.14: A transitive tournament

An orientation of a graph is transitive if xy and yz imply xz. A comparability graph is a graph that allows a transitive orientation of its edges.

WE FIRST SHOW that it is sufficient to prove that tournaments of cutwidth at most k are well - quasi ordered.

Lemma 4.149. *Assume that for every* $k \in \mathbb{N}$ *the class of tournaments of cutwidth* k *is well - quasi -ordered by strong immersions. This implies that the class of all tournaments is well - quasi ordered by strong immersions.*

Proof. By means of contradiction; let (T_i) be a sequence of tournaments such that no T_i strongly immerses in T_j whenever $i < j$. Let $\mathcal{T} = \{T_i\}$ and let

$$S \quad = \quad \mathcal{T} \setminus \{T_1\}.$$

Then there is a digraph that does not strongly immerse in any tournament of S — namely — T_1.

By Theorem 4.147 all tournaments of S have cutwidth at most k (for some $k \in \mathbb{N}$). This contradicts the assumption. $\square$

Linked layouts

Let G be a digraph and let

$$\mu \quad = \quad x_1 \quad \cdots \quad x_n$$

be a layout of G. Write $B_i = \{x_1, \cdots, x_i\}$, $A_i = \{x_{i+1}, \cdots, x_n\}$ and let F_i be the set of edges with tail in B_i and head in A_i.

The layout μ is <u>linked</u> if for all $h < j$ such that $|F_h| = F_j| = t$ and for all $h \leqslant i \leqslant j$ $|F_i| \geqslant t$ there are t edge - disjoint directed paths from B_h to A_j.

Lemma 4.150. *Let G be a digraph of cutwidth k. There is a linked layout of $\hat{G}$ of cutwidth k.*

Proof. For a layout μ of cutwidth at most k let

$$n_s = |\{i \mid |F_i| = s\}|.$$

Choose a layout μ of cutwidth at most k such that the sequence $(n_0, n_1, \cdots)$ is lexicographically as large as possible.

Assume that this layout μ is not linked. Then there exist $h < j$ with $F_h| = |F_j| = t$ and for all $h \leqslant i \leqslant j$ $|F_i| \geqslant t$ and there are not t edge-disjoint paths from B_h to A_j. By Menger's theorem [119] there exists a partition $\{P, Q\}$ of $V(G)$ with $B_h \subseteq P$ and $A_j \subseteq Q$ and there are less than t arcs from P to Q. Let F be the set of arcs with tail in P and head in Q. Choose the partition $\{P, Q\}$ such that $|F|$ is as small as possible.

Let $p = |P|$ and let

$$\mu' = x_1' \cdots x_p' \ x_{p+1}' \cdots x_n'$$

be the layout that puts all elements of P before the elements of Q and that keeps the ordering within the parts P and Q the same as in μ.

We <u>claim</u> that μ' has cutwidth k. We first show that $|F_i'| \leqslant k$ for all $i \neq p$ (where F_i' is the set of edges with tail in $B_i' = \{x_1', \cdots, x_i'\}$ and head in $A_i' = \{x_{i+1}', \cdots, x_n'\}$).

To see that let $i < p$ and choose r such that

$$B_i' = B_r \cap P \quad \text{and} \quad A_i' = A_r \cup Q.$$

Then $r < j$ and $A_j \cap (B_r \cup P) = \emptyset$.

Since we chose $|F|$ minimal and since $B_h \subseteq P \subseteq B_r \cup P$ we have that $|N^+(B_r \cup P)| \geqslant |F|$. [120]

We have [121]

[119] Max flow $=$ min cut

[120] $N^+(S)$ is the set of edges with tail in S and head in $V \setminus S$.

[121] We use the fact that $|N^+(S)|$ ($S \subseteq V$) is a <u>submodular function</u> — that is — for any sets S and T: $|N^+(S)| + |N^+(T)| \geqslant |N^+(S \cup T)| + |N^+(S \cap T)|$. This property is also known as the principle of diminishing returns.

$$|N^+(B_r)| + |N^+(P)| \quad \geqslant \quad |N^+(B_r \cap P)| + |N^+(B_r \cup P)| \quad \text{that is}$$
$$|F_r| + |F| \quad \geqslant \quad |F_i'| + |N^+(B_r \cup P)| \quad \geqslant \quad |F_i'| + |F|$$

and this implies

$$|F_i'| \quad \leqslant \quad |F_r|.$$

It follows that $|F_i'| \leqslant k$ for $i < p$ and similarly $|F_i'| \leqslant k$ for $i > p$. Since $F_p' = F$ and $|F| \leqslant k$ this proves that μ' has cutwidth k.

We $\underline{\text{claim}}$ that $n_s' \geqslant n_s$ for $s < t$. Let $0 \leqslant s \leqslant t-1$ and let r be such that $|F_r| = s$. We actually have that $F_r' = F_r$. To see that observe that we must have $r < h$ or $r > j$. Assume $r < h$. Then $B_r \subseteq P$ and so $B_r' = B_r$ and $F_r' = F_r$.

This shows that $n_s' = n_s$ for all $s < t$ and

$$|F_r'| \quad < \quad t \quad \Rightarrow \quad r \quad < \quad h \quad \text{or} \quad r \quad > \quad j$$

WE ARRIVED AT A CONTRADICTION since $|F_p'| < t$ and $h \leqslant p \leqslant j$. Therefore μ is linked.

This proves the lemma. $\square$

Gap sequences

Let $(Q, \ll)$ be a quasi - order and let $k \in \mathbb{N}$. A (Q, k) - gap sequence is a triple (P, f, a) where

- P is a path

- f is a map $V(P) \to Q$

- a is a map $E(P) \to \{0, \cdots, k\}$.

Define a quasi - order on (Q, k) - gap sequences as follows. For two (Q, k) - gap sequences (P, f, a) and (R, g, b) let $(P, f, a) \preceq (R, g, b)$ if

$$P \quad = \quad p_1 \quad \cdots \quad p_m \quad \text{and} \quad R \quad = \quad r_1 \quad \cdots \quad r_n$$

and there exists a map

$$1 \quad \leqslant \quad s(1) \quad < \quad s(2) \quad < \quad \cdots \quad < s(m) \quad \leqslant \quad n$$

such that

- for all i: $f(p_i) \ll g(r_{s(i)})$

- for all i: if $e = p_i p_{i+1}$ then $a(e) \leqslant b(e')$ for all $e' \in E(R)$ that are on the path from $r_{s(i)}$ to $r_{s(i+1)}$.

By Theorem 4.101 when $(Q, \ll)$ is a well - quasi order then $\preceq$ is a well - quasi order on (Q, k) - gap sequences.

Marches

A <u>march</u> μ is a sequence

$$e_1 \quad \cdots \quad e_k$$

of elements. The set $\{e_1, \cdots, e_k\}$ is the support of μ and $k = |\mu|$ is the length of μ. We write $e_i = \mu(i)$.

Define an equivalence on pairs of marches as follows. Two pairs pairs of marches (μ_1, v_1) and (μ_2, v_2) are equivalent if

- $|\mu_1| = |\mu_2|$

- $|v_1| = |v_2|$

- for all i and j $\mu_1(i) = v_1(j) \Leftrightarrow \mu_2(i) = v_2(j)$.

Codewords

A <u>codeword</u> of type k is a pair (P, f) where

- P is a path say $P = p_1 \cdots p_n$

- f is a function with domain $V(P)$ which maps a vertex p_i of P to a pair of marches (μ_i, v_i) both of length at most k such that

  - $|v_i| = |\mu_{i+1}|$
  - $|\mu_1| = |v_n| = 0$.

The cutsize function $a : E(P) \rightarrow \{0, \cdots, k\}$ maps each edge (p_i, p_{i+1}) to $|v_i| = |\mu_{i+1}|$.

Let $\mathcal{C}_k$ be the set of all codewords of type k. Define a quasi - order on $\mathcal{C}_k$ as follows. Let (P, f) and (R, g) be two codewords of type k and let a and b be their cutsize functions. Then (P, f, a)

and (R, g, b) are (Q, k) - gap sequences where Q is the set of pairs of marches ordered by equivalence. Let $(P, f) \preceq (R, g)$ if $(P, f, a) \preceq (R, g, b)$.

Lemma 4.151. *For each* k *$(\mathcal{C}_k, \preceq)$ is well - quasi ordered.*

Proof. The set Q of pairs of marches of length at most k is well - quasi ordered by equivalence because there are only a finite number of equivalence classes.   □

Let G be a tournament of cutwidth k and let

$$x_1 \quad \cdots \quad x_n$$

be a linked layout of G of cutwidth k. Define B_i, A_i and F_i as before. We have that for all $h < j$ that satisfy

$$|F_h| \;=\; |F_j| \;=\; t \quad \text{and} \quad \forall_{h \leqslant i \leqslant j} \; |F_i| \;\geqslant\; t \quad (4.20)$$

there are t edge disjoint paths $P_1, \cdots, P_t$ from B_h to A_j.

The following lemma makes sure that we can find marches with support F_i such that the sth elements of them are edges of P_s.

Lemma 4.152. *There exist marches* μ_i *with support* F_i *such that all* $h < j$ *that satisfy* (4.20) *there are edge - disjoint paths* $P_1, \cdots, P_t$ *such that for* $s \in [t]$ *the* s^{th} *term of* μ_h *and the* s^{th} *term of* μ_j *are edges of* P_s.

Exercise 4.96

Prove Lemma 4.152.

HINT: Fix t. Let $i(1) < \cdots < i(\ell)$ be the indices i with $|F_i| = t$. For $j = 1, \cdots, \ell$ choose the march $\mu_{i(j)}$ (ie choose a linear order of the elements of $F_{i(j)}$) such that the sth element extends the path P_s.

Encoding

Let G be a tournament and let $g_1 \cdots g_n$ be a layout of G with cutwidth k. Map G to a codeword of type k as follows.

Let G denote the path with vertices $g_1, \cdots, g_n$. Define $\mu_0 = \mu_n = \varnothing$ and let μ_i be a march as in Lemma 4.152. Let g be the map

We use the same symbol G for the tournament and its layout.

$$g(g_i) \quad = \quad (\mu_{i-1}, \mu_i)$$

for $i \in [n]$. Then (G, g) is a codeword of type k.

Lemma 4.153. *Let G and H be tournaments of cutwidth k and let (G, g) and (H, h) be codewords of G and H. Assume $(H, h) \preceq (G, g)$ in $(\mathcal{C}_k, \preceq)$. Then H immerses strongly in G.*

Proof. There are linked layouts of the tournaments G and H that give rise to the codewords (G, g) and (H, h) — say —

table of notations:

	H	G
	$(h_j)_1^m$	$(g_i)_1^n$
	D_j, C_j, F_j	B_i, A_i, E_i
	marches: ν_j	marches: μ_i

$$G \quad = \quad g_1 \quad \cdots \quad g_n \qquad \text{and} \qquad H \quad = \quad h_1 \quad \cdots \quad h_m.$$

For the layout (g_i) define B_i and A_i as above and let E_i be the set of edges with tail in B_i and head in A_i. Similarly define D_j, C_j and F_j for the layout (h_j) of H. Denote the cutsize functions of (g_i) and (h_j) as b and a.

Let the μ_i be marches with support E_i as in Lemma 4.152 and — similarly — let the ν_i be marches with support F_i.

We have that $(H, h) \preceq (G, g)$ which implies there are

$$1 \quad \leqslant \quad r(1) \quad < \quad r(2) \quad < \quad \cdots \quad < \quad r(m) \quad \leqslant \quad n$$

such that

- $h(h_i)$ and $g(g_{r(i)})$ are equivalent pairs of marches

- if $e = h_i h_{i+1}$ then $a(e) \leqslant b(e')$ for every edge e' on the path $g_{r(i)} \rightsquigarrow g_{r(i+1)}$.

Since $h(h_i) = (\nu_{i-1}, \nu_i)$ and $g(g_{r(i)}) = (\mu_{r(i)-1}, \mu_{r(i)})$ are equivalent pairs of marches we have that

$$|F_i| = |E_{r(i)}|, \quad |F_{i-1}| = |E_{r(i)-1}| \quad \text{and} \quad |F_{i-1} \cap F_i| = |E_{r(i)-1} \cap E_{r(i)}|$$

The second property implies

$$|E_{r(i)}| = |E_{r(i+1)-1}| = |F_i| \quad \text{and}$$

$$|E_j| \geqslant |F_i| \quad \text{for all } r(i) \leqslant j \leqslant r(i+1) - 1.$$

Let $1 \leqslant i \leqslant m$. For $e \in F_i$ there are directed paths $P_i(e)$ in G with the following properties.

(a) the paths in $\{P_i(e) \,|\, e \in F_i\}$ are pairwise edge - disjoint

(b) the first edge of $P_i(e)$ is in $E_{r(i)}$ and it has tail $g_{r(i)}$ if and only if e has tail h_i

(c) the last edge of $P_i(e)$ is in $E_{r(i+1)}$ and it has head $g_{r(i+1)}$ if and only if e has head h_{i+1}

(d) all internal vertices of $P_i(e)$ are in $\{g_{r(i)+1}, \cdots, g_{r(i+1)-1}\}$

(e) let e be the sth term of march v_i. The first edge of $P_i(e)$ is the sth term of $\mu_{r(i)}$ and the last edge of $P_i(e)$ is the sth term of $\mu_{r(i+1)-1}$.

Let $e = h_h h_j$ be an edge of H with $h < j$. Then $e \in F_i$ for $h \leqslant i < j$. The reader is invited to check that the appropriate paths $P_i(e)$ glue together — to be precise — let $e \in E(H)$ and let $e = h_h h_j$ for $h < j$. There is a directed path $\eta(e) = g_{r(h)} \rightsquigarrow g_{r(j)}$ in G such that

(f) no vertex of $\{g_{r(1)}, \cdots, g_{r(m)}\}$ is an internal vertex of $\eta(e)$

(g) all the paths in $\{\eta(e) \,|\, e = h_h h_j \text{ where } h < j\}$ are pairwise edge - disjoint

(h) if e is the sth term of v_h then the first edge of $\eta(e)$ is the sth term of $\mu_{r(h)}$

(i) if e is the tth term of v_{j-1} then the last edge of $\eta(e)$ is the tth term of $\mu_{r(j)-1}$.

Let $h < j$ and let $e = h_j h_h$ be an edge of H. Then $g_{r(j)} g_{r(h)}$ is an edge of G. For these edges define $\eta(e)$ as the edge $g_{r(j)} g_{r(h)}$; this is a directed path in G of length one and it is edge - disjoint from the paths $\eta(e)$ for edges in H that point forward in its layout.

Define $\eta(h_i) = g_{r(i)}$. This completes the definition of η which is a strong immersion of H in G.

This proves the lemma.   □

Theorem 4.154. *The class of tournaments is well - quasi ordered by the strong immersion - relation.*

Proof. By Lemma 4.149 it is sufficient to show that a class of tournaments of cutwidth at most k is well - quasi ordered by strong immersions. Let (T_i) be a sequence of tournaments of cutwidth at most k. Their codewords are elements of $\mathcal{C}_k$ and these are well - quasi ordered. — So — there exist $i < j$ such that the codeword of T_i is dominated by the codeword of T_j. By Lemma 4.153 this implies that T_i strongly immerses in T_j. $\sqcap$

Remark 4.155. Orientations of complete bipartite graphs are well - quasi ordered under strong immersions — moreover — the immersion relation respects the parts of the bipartition.

4.14.4 Domination in tournaments

Exercise 4.97

Every acyclic digraph has a unique independent dominating set.

HINT: This result has been attributed to Von Neumann and Morgenstern.

In 2017 Bousquet, Lochet and Thomassé proved the Erdős - Sands - Sauer - Woodrow conjecture. In this section we review their proof.

> There exists a function $g : \mathbb{N} \to \mathbb{N}$ so that if the arcs of a tournament are colored with k colors there is a set S with at most $g(k)$ vertices such that for every vertex x there is a monochromatic path from S to x.

Let T be a tournament and let the arcs be colored with k colors. In order to formulate the conjecture above as a domination problem we would want each color to induce a quasi - order — so — we take the transitive closure of the set of arcs of each color. — Clearly — this may introduce multiple arcs between pairs of vertices.

J. von Neumann and O. Morgenstern, *Theory of games and economic behavior*, Princeton University Press, 1944.

The transive closure of a binary relation $(Q, \preceq)$ is the smallest transitive relation $(Q, \leqslant)$ which contains $(Q, \preceq)$.

A multiset is a set together with a multiplicity function which maps the elements of the set to $\mathbb{N}$. It is sometimes called a bag.

A digraph is an orientation of a graph. So it has no loops, no multiple arcs and no directed cycles of length two.

Definition 4.156. A complete multi digraph is a set D of vertices and a multiset of arcs A such that

1. every arc is an ordered pair of vertices

2. every two vertices form at least one arc

A complete multi digraph can have cycles of length two (but no loops).

For a (multi-) digraph (D, A) and $x \in D$ define the closed in-neighborhood as

A multi digraph is a set of vertices together with a multiset of ordered pairs of distinct vertices. Multi digraphs can have oriented cycles of length two but not of length one.

$$N^-[x] \;\; = \;\; \{x\} \;\; \cup \;\; \{y \mid (y, x) \in A\}.$$

For a set S we write $N^-[S] = \cup_{x \in S} N^-[x]$. Similarly define N^+.

A set S is domination if $N^+[S] = V$. The domination number of the digraph $\gamma(D)$ is the smallest cardinality of a dominating set.

We prove the following theorem. (Clearly this implies the ESSW - conjecture, above.)

Theorem 4.157. *There exists a function* $f : \mathbb{N} \to \mathbb{N}$ *with the following property. Let* T *be a complete multi digraph whose arcs are the union of* k *quasi - orders then* $\gamma(T) \leqslant f(k)$.

The proof of the theorem makes use of two lemmas.

Let T be a complete multi digraph and let the arcs of T be covered with k quasi - orders — say — $(T, \leqslant_i)$, $(i \in [k])$. For $x \in V(T)$ we write $N_i^-[x]$ for the closed in-neighborhood of x in $(T, \leqslant_i)$.

Lemma 4.158. *Let* T *be a complete multi digraph whose set of arcs is the union of* k *quasi - orders. There exists a probability distribution* $w : V(T) \to [0, 1]$ *and a partition* $\{T_1, \cdots, T_k\}$ *of* $V(T)$ *such that for each* $x \in T_i$

$$w(N^-[x]) \;\; \geqslant \;\; \frac{1}{2k}.$$

Proof. By Theorem 4.136 (on Page 241) there is a probability distribution $w : V(T) \to [0, 1]$ such that $w(N^-[x]) \geqslant 1/2$ for all $x \in V(T)$.

Define for $i \in [k]$

$$T_i \quad = \quad \{\, x \mid w(N_i^-[x]) \;\geqslant\; 1/2k \,\}.$$

Then $\cup T_i = V(T)$.

— Clearly — the sets T_i can be reduced so that the result forms a partition of $V(T)$.

This proves the lemma. $\square$

Let $(P, \preceq)$ be a quasi - order. We identify $(P, \preceq)$ with a digraph (P, A) where the set of arcs is the set of ordered pairs xy with $x \preceq y$ and $x \neq y$. [122]

[122] So $N^-[x]$ is well-defined on the elements of a quasi - order.

Definition 4.159. Let $(P, \preceq)$ be a quasi - order. A set $A \subseteq P$ is $\underline{\varepsilon\text{-dense}}$ in P if there is a probability distribution w on P which satisfies

$$\forall_{x \in A} \quad w(N^-[x]) \;\geqslant\; \varepsilon.$$

Lemma 4.160. *There exists a function* $g : [0,1] \to \mathbb{N}$ *with the following property. In every quasi - order* $(P, \preceq)$ *if* $C \subseteq B$ *are subsets of* P *such that* B *is* ε-*dense in* P *and* C *is* ε-*dense in* B *then there exists a set of* $g(\varepsilon)$ *elements of* P *that dominate* C.

Proof. Let $w : P \to [0,1]$ and $w_B : B \to [0,1]$ be probability distributions that show that B is ε-dense in P and C is ε-dense in B — that is —

$$\forall_{x \in B} \quad w(N^-[x]) \;\geqslant\; \varepsilon \quad \text{and} \quad \forall_{x \in C} \quad w_B(N^-([x]) \;\geqslant\; \varepsilon.$$

Define the function $g(\varepsilon) = \left\lfloor \frac{\ln(\varepsilon)}{\ln(1-\varepsilon)} \right\rfloor + 1$.

Select —at random and according to probability distribution w — a multiset S of $g(\varepsilon)$ elements of P. Then

$$\forall_{x \in B} \quad \mathbb{P}(x \in N^+[S]) \;\geqslant\; 1 - (1-\varepsilon)^{g(\varepsilon)} \;>\; 1 - \varepsilon.$$

By linearity of expectation of w_B there exists a set S such that $w_B(N^+[S]) > 1 - \varepsilon$.

Since $w_B(N^-[x]) \geqslant \varepsilon$ for all $x \in C$ we have that $N^-[x]$ intersects $N^+[S]$ for all $x \in C$. By transitivity this implies that S dominates C.

This proves the lemma.   □

We now present the proof of Theorem 4.157.

Theorem. *Let* $k \in \mathbb{N}$ *and let* T *be a complete multi digraph whose arcs are the union of* k *quasi - orders then* $\gamma(T) = O(k^{k+2} \cdot \ln(2k))$.

Proof. Let $P_1 = \{T_1, \cdots, T_k\}$ be a partition of $V(T)$ as mentioned in Lemma 4.158. Repeat this partitioning process $k + 1$ times to obtain a sequence of partitions $P_1, \cdots, P_{k+1}$ which we specify as

$$P_i \quad = \quad \{T_{j_1 \cdots j_i} \mid j_1, \cdots, j_i \in [k]\}$$

so that for each $\ell \leqslant k + 1$ $T_{j_1 \cdots j_\ell}$ is a subset of $T_{j_1 \cdots j_{\ell-1}}$.

Let $w_{j_1 \cdots j_{\ell-1}}$ be a probability distribution (as in Lemma 4.158) such that

$$w_{j_1 \cdots j_{\ell-1}}(N^-_{j_\ell}[x]) \quad \geqslant \quad \frac{1}{2k}$$

for all $x \in T_{j_1 \cdots j_\ell}$.

In this formula $N^-_{j_\ell}([x])$ denotes the closed in-neighborhood of x in the j^{th}_ℓ quasi order.

BY THE PIGEONHOLE PRINCIPLE every sequence $j_1 \cdots j_{k+1}$ in $[k]^{k+1}$ contains $i < \ell$ such that $j_i = j_\ell$. Apply Lemma 4.160 with

$$P = T_{j_1 \cdots j_{\ell-1}} \quad B = T_{j_1 \cdots j_i} \quad \text{and} \quad C = T_{j_1 \cdots j_\ell}.$$

It follows that there exists a set of at most $g(1/2k)$ elements that dominates $T_{j_1 \cdots j_\ell}$ and so it dominates $T_{j_1 \cdots j_{k+1}}$.

We can conclude that $\gamma(T) \leqslant k^{k+1} \cdot g(1/2k)$. Notice that $g(1/2k) \leqslant \ln(2k) \cdot (2k - 1/2 + o(1))$ — that is — $\gamma(T) = O(k^{k+2} \cdot \ln(2k))$.

This proves the theorem.   □

AS FAR AS WE KNOW the following conjecture is open. (It was posed by Sanders, Sauer and Woodrow in 1982.)

Conjecture 4.161. *There exists a function* $f \in \mathbb{N}^{\mathbb{N}}$ *with the following property. Let* D *be a multi digraph whose set of arcs is a union of* k *quasi - orders. Then* D *has a dominating set which is the union of* $f(k)$ *independent sets.*

4.15 Immersions

In this chapter graphs and digraphs are allowed to have multiple edges but no loops unless stated otherwise.

Robertson and Seymour proved that the class of all graphs is well quasi - ordered by weak immersions. Whether the same holds true for strong immersions is an open problem. [123]

In this chapter we show that digraphs without k - alternating paths are well quasi - ordered by strong immersions.

[123] See Page 246 for the definitions of immersions.

Exercise 4.98

A graph is 'subcubic' if every vertex has degree at most 3. Show that the class of subcubic graphs is well quasi - ordered by strong immersions. — Also — for subcubic graphs H is a topological minor of G if and only if it is a minor.
HINT: Let G and H be subcubic. Show that H immerses in G if and only if H is a minor of G.

Chun-Hung Liu and Irene Muzi show that digraphs without k-alternating paths are well quasi - ordered by strong immersions. Before we take a closer look at their proof let us take some time off to meditate on an important result on topological minors.

Chun-Hung Liu and Irene Muzi, Well - quasi - ordering digraphs with no long alternating paths by the strong immersion relation. Manuscript on arXiv: 2007.15822, 2020.

4.15.1 Intermezzo: Topological minors

Relax: we present only facts; no proofs; just try to understand what's going on...

Definition 4.162. Let G and H be graphs. The graph H is a topological minor of G if some subgraph of G is isomorphic to a subdivision of H.

A graph H is a topological minor of G if there is a <u>homeomorphic</u> embedding of H in G — that is — a map $\eta : H \to G$ such that

1. the map $\eta : V(H) \to V(G)$ is injective

2. η maps each edge $\{x, y\} \in E(H)$ to a path $\eta(x) \rightsquigarrow \eta(y)$ in G such that distinct edges of H map to paths in G that have no vertices in common other than endpoints.

Exercise 4.99

Show that K_5 is a minor of the Peterson graph but that it is not a topological minor.

Remark 4.163. Grohe, Marx, Wollan, and Kawarabashi show that finding a topological minor is fixed parameter - tractable: there exists a cubic algorithm to test if a graph H of 'constant size' is a topological minor of a graph G.

Graphs are <u>not</u> well quasi - ordered by topological minors. — To see that — let P_i be a path with i vertices and construct a graph G_i as follows.

- duplicate every edge of P_i

- attach two new vertices to each end of P_i.

The sequence (G_i) is an infinite antichain in the topological minor order.

Chun-Hung Liu and Robin Thomas prove that this is the only obstruction.

Definition 4.164. A Robertson chain of length k is a graph obtained from a path of length k by duplicating each edge.

Theorem 4.165. *Let* $k \in \mathbb{N}$ *and let* $(Q, \leqslant_Q)$ *be a well quasi - order. Let* (G_i) *be a sequence of graphs without Robertson chain of length* k *and let* $\phi_i : V(G) \to Q$ *be a labeling of the vertices of* G_i *with elements of* Q*. There exist* $j < j'$ *and a homeomorphism* $\eta : G_j \to G_{j'}$ *which satisfies*

$$\forall_{x \in V(G_j)} \qquad \phi_j(x) \quad \leqslant_Q \quad \phi_{j'}(\eta(x)).$$

For subcubic graphs the topological minor relation is equivalent with the minor relation. Let G be an arbitrary graph. We can map it to a subcubic graph G' as follows. Replace a vertex x by a cycle with $d(x)$ vertices. Each vertex in the cycle receives one neighbor of x as a neighbor outside the cycle. For what classes of graphs holds $G' \leqslant_{\text{top minor}} H' \Rightarrow G \leqslant_{\text{top minor}} H$?

Further reading on topological minors:

C.-H. Liu, *Graph structures and well-quasi-ordering*, PhD dissertation, Georgia Institute of Technology, 2014.

C.-H. Liu and R. Thomas, Robertson's conjecture I. Well - quasi - ordering bounded treewidth graphs by the topological minor relation. Manuscript on arXiv: 2006.00192, 2020.

M. Grohe, D. Marx, K. Kawarabashi and P. Wollan, Finding topological subgraphs is fixed parameter tractable. Manuscript on arXiv: 1011.1827, 2010.

4.15.2 *Strong immersions in series - parallel digraphs*

A <u>thread</u> is a digraph whose underlying graph is a path. A thread P in a digraph D is <u>k-alternating</u> if it changes direction k times — that is — if it has k vertices that have in-degree in P equal to 0 or out-degree in P equal to 0.

Chun-Hung Liu and Irene Muzi prove the following theorem.

Theorem 4.166. *Let* $k \in \mathbb{N}$ *and let* (D_i) *be a sequence of digraphs without* k*-alternating thread. Let* $(Q, \leqslant)$ *be a well quasi - order and for all* i *let* $\phi_i : V(D_i) \to Q$. *There exist* $j < j'$ *and a strong immersion* η *of* D_j *into* $D_{j'}$ *such that for all* $x \in V(D_j)$

$$\phi_j(x) \quad \leqslant \quad \phi_{j'}(\eta(x)).$$

In this chapter we take a close look at the proof of this theorem. The proof is by induction on k. We present the <u>base case</u> $k = 1$ as an exercise.

Exercise 4.100

Let D be a digraph in which every thread is a directed path. Then D is obtained from a directed path or cycle (possibly of length two) by multiplying edges.

Exercise 4.101

Let (D_i) be a sequence of digraphs without 1-alternating thread. Let $(Q, \leqslant)$ be a well quasi - order and let $\phi_i : V(D_i) \to Q$. There exist $j < j'$ such that there is a strong immersion η of D_j in $D_{j'}$ which satisfies

$$\forall_{x \in V(D_j)} \qquad \phi_j(x) \quad \leqslant \quad \phi_{j'}(\eta(x)).$$

Hint: Use the gap theorem.

Perhaps we should do this in class... Work this out in detail!

STEP NUMBER TWO is a proof of the fact that one - way series - parallel triples are well quasi - ordered by strong immersions.

Definition 4.167. A triple (D, s, t) is a two - terminal graph if D is a multigraph and $s, t \in V(D)$ and either

- $V(D) = \{s, t\}$ and $E(D) = \{\{s, t\}\}$

- D is a series composition: there exist two - terminal graphs (D_1, s_1, t_1) and (D_2, s_2, t_2) and $s = s_1$ and $t = t_2$ and D is obtained from the union of D_1 and D_2 by identifying t_1 and s_2

- D is a parallel composition: (D, s, t) is obtained from a union of two - terminal graphs (D_1, s_1, t_1) and (D_2, s_2, t_2) by identifying $s = s_1 = s_2$ and $t = t_1 = t_2$.

A biconnected multigraph is a two - terminal graph if and only if it is confluent. That is, for any two edges every cycle that contain them, meets the endpoints in the same relative order. A multigraph is confluent if and only if it contains no subgraph which is a subdivision of K_4 — so — its underlying simple graph has treewidth two.

The two vertices s and t are the 'terminals' of the graph.

4.15.3   Intermezzo on 2 - trees

The underlying simple graph of a 2 - terminal graph is a partial 2 - tree — that is — it is a subgraph of a 2 - tree.

To define a 2 - tree: any graph that is an edge is a 2-tree. When T is a 2-tree and t a triangle then a new 2-tree is obtained from the disjoint union by identifying the endpoints of an edge in T with the endpoints of an edge in t.

The partial 2 - trees are the graphs of treewidth 2. They are the graphs that do not have K_4 as a minor.

Labeled and unlabeled biconnected partial two-trees can be enumerated (like trees). The enumeration of the 'rooted' graphs (where the root is a pair $\{s, t\}$) serves as a first step. See Chapter 4 in: Ton Kloks, "Treewidth." PhD Thesis, 1993.

'To understand what the elements of a combinatorial structure look like you should try to enumerate them.' (De Bruijn.)

     . . .

Figure 4.15: Enumeration of 2-trees

Exercise 4.102

Any 2-tree has an orientation which is acyclic.

Let G be a biconnected graph of treewidth two. A cell-completion of G is obtained from G as follows. Let s and t be nonadjacent vertices in G. If $G - s - t$ has at least three components then add an edge $\{s, t\}$ in the cell-completion.

> When G is biconnected and has treewidth two then its cell-completion is unique and it is a tree of cycles.

A tree of cycles is a graph defined recursively, as follows.

(i) any graph that is a cycle is a tree of cycles

(ii) Let C be a cycle and let T be a tree of cycles. Then another tree of cycles is obtained from the union by identifying the endpoints of an edge in C with the endpoints of an edge in T.

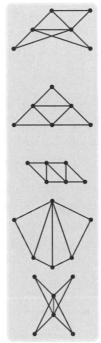

Figure 4.16: A clip from the cover

4.15.4 Series parallel - triples

When a two-terminal graph is not biconnected then its cutvertices and blocks form a path: every cutvertex is in two blocks, every

block is incident with at most two cutvertices, and there are two
blocks that are incident with exactly one cutvertex.

The underlying simple graph of a two - terminal graph has
treewidth two — that is — a graph without a subgraph homeomorphic to K_4. However, notice that the claw can not be generated
as a two - terminal graph: every cutvertex in a two - terminal graph
separates the graph in two components; one contains s and the
other contains t.
A graph is the underlying graph of a two - terminal graph if and
only if it is a graph of treewidth two of which the cutvertices and
blocks form a path. Every block has a minimal triangulation (into
a 2-tree) in which the two cutvertices (including s and t) form an
edge.

The Figure 4.17 shows a minimal triangulation of a 2-terminal
graph. To specify the 2-terminal graph each edge of this minimal
triangulation is labeled with a multiplicity; ie an element of $\in$
$\mathbb{N} \cup \{0\}$. (The multiplicity - labels are not shown.)

The only edges in a minimal
triangulation of a block that
can have multiplicity zero
are edges that are minimal
separators.

Since a 2-terminal graph has treewidth 2 each block in a minimal
triangulation is a 2-tree. It has a coloring with three colors such
that every pair of colors induces a tree. Similarly, a 2-tree has a
3-partition of its edges such that each part is a tree.

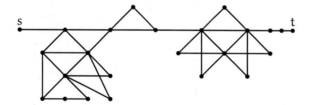

Figure 4.17: The figure
shows a minimal triangulations of a 2-terminal graph.

A one-way series-parallel digraph is an orientation of a 2-terminal
graph such that all threads that run from s to t are directed paths.

Definition 4.168. A series - parallel triple (D, s, t) is a directed
graph D whose underlying graph is connected and s and t are

distinct vertices of D such that every thread with ends s and t is a directed path and every cutvertex separates s and t.

A series - parallel triple is <u>one - way</u> if every s, t - thread is a directed path from s to t or if every s, t - thread is a directed path from t to s.

The proof of the following lemma is an easy exercise.

Lemma 4.169. *A series - parallel triple is an orientation of a two - terminal graph* (D, s, t) *such that every thread with ends s and t is a directed path.*

Definition 4.170. Let $(Q, \leqslant)$ be a well quasi - order. Let (D_i, s_i, t_i) (for $i \in \{1, 2\}$) be two series - parallel triples and let $\phi_i : V(D_i) \to Q$. The pair (D_2, ϕ_2) <u>simulates</u> (D_1, ϕ_1) if there exists a strong immersion $\eta : D_1 \to D_2$ which satisfies

$$\eta(s_1) = s_2 \quad \text{and} \quad \eta(t_1) = t_2 \quad \text{and}$$
$$\forall_{x \in V(D_1)} \quad \phi_1(x) \quad \leqslant \quad \phi_2(\eta(x))$$

Definition 4.171. A collection $\mathcal{F}$ of series parallel triples is <u>well - simulated</u> if for every well quasi - order $(Q, \leqslant)$ in any sequence $((D_i, \phi_i))$ of Q-labeled elements of $\mathcal{F}$ there exist $j < j'$ such that $(D_{j'}, \phi_{j'})$ simulates (D_j, ϕ_j).

Parallel compositions

Lemma 4.172. *Let $\mathcal{F}$ be a set of well - simulated one way series parallel triples. Let $\mathcal{F}^p$ be the set of parallel compositions of elements of $\mathcal{F}$. Then $\mathcal{F}^p$ is well - simulated.*

Proof. Let $(Q, \leqslant)$ be a well quasi - order and let (D_i, ϕ_i) be a sequence of Q-labeled series - parallel triples in $\mathcal{F}^p$. By assumption each D_i is a parallel composition of a collection of — say ℓ_i ($\ell_i \in \mathbb{N}$) — series - parallel triples that are in $\mathcal{F}$:

$$D_i \quad \text{is a parallel composition of} \quad \{D_{i,j} \mid j \in [\ell_i]\}$$

The minimal triangulations of the underlying simple graphs in a parallel composition are obtained by gluing 2-trees together along their root - edges $\{s, t\}$. (See Figure 4.17.) A parallel composition encodes as a 'Higman - word' over an alphabet formed by the constituents of the composition.

Define a <u>word</u> a_i — which encode (D_i, ϕ_i) — as the sequence

$$a_i \quad = \quad (D_{i,1}, \psi_1) \quad \cdots \quad (D_{i,\ell_i}, \psi_{\ell_i})$$

where the ψ_j in this formula are simply the restrictions of ϕ_i to $V(D_{i,j})$ (for $j \in [\ell_i]$).

It now follows from Higman's Lemma that there exist $j < j'$ such that $(D_{j'}, \phi_{j'})$ simulates (D_j, ϕ_j).

This proves the lemma. $\square$

$\mathcal{F}$ - Series parallel trees

Series compositions are not easy to deal with since immersions may 'stretch out' the domain. To handle this we introduce series - parallel trees.

Rooted digraphs (D, r) are digraphs with a root. We let strong immersions of rooted digraphs preserve the root.

Definition 4.173. A set of rooted digraphs is <u>well - behaved</u> if for any well quasi - order $(Q, \leqslant)$ and any sequence in the set of rooted digraphs (D_i, r_i) with a labeling $\phi_i : V(D_i) \to Q$ there exist $j < j'$ such that there is a strong immersion η of (D_j, r_j) in $(D_{j'}, r_{j'})$ which satisfies $\eta(r_j) = r_{j'}$ and

$$\forall_{x \in V(D_j)} \quad \phi_j(x) \quad \leqslant \quad \phi_{j'}(\eta(x))$$

Figure 4.17 suggests an encoding of one way series parallel triple as a 'word' over an alphabet which is the set of blocks and to use Higman's lemma - with - a - gap. The alphabet (set of blocks) is well quasi - ordered by homeomorphic embedding.

Let (D, r) be a rooted digraph. Associate with (D, r) a rooted tree T of which the nodes are the cutvertices (including r) and the blocks of D. A block and a cutvertex are adjacent in T when the block contains the cutvertex. The root of T maps to the root of the digraph.

Definition 4.174. Let $\mathcal{F}$ be a set of rooted digraphs. A rooted digraph (D, r) is an $\mathcal{F}$ <u>series parallel tree</u> if

1. the block that contains r is in $\mathcal{F}$

2. If B is a block and c is the cutvertex that separates it from its parent then $(B, c) \in \mathcal{F}$

3. every thread from r to a cutvertex is a directed path

4. every block contains at most two cutvertices of D; so   every block B which has a child - block is a series parallel triple. [124]

Truncations and portraits

Let (B, x, y) be a middle block of a series parallel tree.   Let $\{X, Y\}$ be a partition of $V(B)$ such that $x \in X$, $y \in Y$ and the number of edges with one end in X and the other end in Y is equal to the maximal number of edge - disjoint threads between x and y.   A truncation is the series parallel triple obtained by <u>shrinking</u> one of the two parts X or Y to one vertex.

A block is a middle block if it has two cutvertices.

$\{X, Y\}$ is a minimum cut.

In an $\mathcal{F}$ - series parallel tree add the two truncations of every middle block to the tree; by subdividing the two edges incident with the middle block. The new trees are called portraits.

The gap - theorem — applied to these portraits — proves the following lemma. (We omit the proof.)

Lemma 4.175.

Let $\mathcal{F}$ be a set of rooted digraphs which behaves well

- *$\mathcal{F}'$ is the set of series parallel triples (D, s, t) with $(D, s) \in \mathcal{F}$ and $t \in V(D - s)$*

- *$\mathcal{F}''$ is the set of truncations of elements of $\mathcal{F}'$.*

If $\mathcal{F}'$ and $\mathcal{F}''$ are well - simulated then the set of $\mathcal{F}$ - series parallel trees behaves well.

Let $\mathcal{F}$ be a set of one way series parallel - triples and assume that $\mathcal{F}$ is well - simulated. Let $\mathcal{F}^s$ denote the set of all series extensions of elements of $\mathcal{F}$. The following exercise initiates a proof to show that $\mathcal{F}^s$ is well - simulated.

The series extensions of $\mathcal{F}$ is the set $\mathcal{F}^*$ of which the elements in a word are chained by identifying t_i and s_{i+1}.

Exercise 4.103

Let $\mathcal{F}$ be a set of one way series parallel triples which is well - simulated;

- $\mathcal{F}^s$ is the set of one way series parallel triples that are series extensions of elements of $\mathcal{F}$

- $\mathcal{F}^t$ is the set of all truncations of elements of $\mathcal{F}$.

If $\mathcal{F}^t$ is well - simulated then $\mathcal{F}^s$ is well - simulated.

Hint: Use Lemma 4.175.

4.15.5   A well quasi - order for one way series parallel - triples

In this section let $\mathcal{F}$ be a set of one way series parallel triples. For $k \in \mathbb{N}$ let $\mathcal{F}^k \subseteq \mathcal{F}$ be the set of one way series parallel triples that do not contain a k-alternating path.

Exercise 4.104

Let (D, s, t) be a one way series parallel triple. Design an algorithm to calculate the maximal number $k \in \mathbb{N}$ for which (D, s, t) has a k-alternating path.

Chun-Hung Liu and Irene Muzi prove the following lemma.

Lemma 4.176. *Let $(Q, \leqslant)$ be a well quasi - order. Let $((D_i, s_i, t_i))$ be a sequence in $\mathcal{F}^k$ and let $\phi_i : V(D_i) \to Q$. There exist $j < j'$ and a strong immersion $\eta : D_j \to D_{j'}$ such that $\eta(s_j) = s_{j'}$ and $\eta(t_j) = t_{j'}$ and*

$$\forall_{x \in V(D_j)} \qquad \phi_j(x) \quad \leqslant \quad \phi_{j'}(\eta(x))$$

Proof. Cover the set $\mathcal{F}$ with the following collections of one way series parallel triples.

1. A_0 is the set of series parallel triples that consist of one edge

2. $A_{0,0} = A_0$

For k and i in $\mathbb{N} \cup \{0\}$ define

3. $A_{k,2i+1}$ is the set of all parallel extensions of elements in $A_{k,2i}$

4. $A_{k,2i+2}$ is the set of all series extensions of elements in $A_{k,2i+1}$

5. $A_{k+1} \subseteq \mathcal{F}$ is the set of one way series parallel triples that have no $(k+1)$ - alternating path that starts in s or t

6. $A_{k+1,0}$ is the set of elements in $\mathcal{F}$ such that either

   - every $(k+1)$ - alternating path with s on one end contains t and there is no $(k+1)$ - alternating path with t on one end

   or

   - every $(k+1)$ - alternating path with t on one end contains s and there is no $(k+1)$ - alternating path with s on one end.

The lemma is proved in the following steps.

(a) every series - irreducible triple in A_{k+1} is in $A_{k,3}$

(b) $A_{k+1} \subseteq A_{k,4}$

(c) for $(D, s, t) \in \mathcal{F}$:

   - if $(D, s, t) \in A_k$ then every truncation[125] is in A_k
   - if $(D, s, t) \in A_{k,0}$ then every truncation is in $A_{k,0}$.

[125] with respect to a partition $\{S, T\}$ with $s \in S$ and $t \in T$ and a minimal number of crossing edges.

The next claim is proved via Lemma 4.175.

(d) if A_k is well - simulated then $A_{k,0}$ is well - simulated

(e) let $k > 0$ and $\ell \geqslant 0$. All truncations of elements of $A_{k,\ell}$ are elements of $A_{k,\ell}$.

(f) for $k, \ell \geqslant 0$ if $A_{k,0}$ is well - simulated then $A_{k,\ell}$ is well - simulated.

It now easily follows by induction on k that A_k is well - simulated: This is clearly true for $k = 0$. When A_{k-1} is well - simulated then by (d) $A_{k-1,0}$ is well - simlulated. By (f) $A_{k-1,4}$ is well - simulated and since $A_k \subseteq A_{k-1,4}$ (by (b)) A_k is well - simulated.

This proves the lemma since — obviously — $\mathcal{F}^k \subseteq A_k$. □

4.15.6   Series parallel separations

Definition 4.177. Let D be a digraph. A <u>separation</u> of D is a pair of <u>edge disjoint</u> subgraphs (A, B) such that $A \cup B = D$. The order of the separation is $|V(A \cap B)|$.

The intersection $A \cap B$ of two digraphs is of course what you think it is.

Definition 4.178. A <u>series parallel separation</u> of a digraph D is a separation (A, B) of D with $V(A \cap B) = \{s, t\}$ and such that (A, s, t) is a one way series parallel triple.

Exercise 4.105

Let G be a graph of <u>treewidth</u> w. Let $\mathcal{S}$ be a collection of subsets of V and assume that $G[S]$ is connected for each $S \in \mathcal{S}$. Let $k \in \mathbb{N}$. One of the two following statements holds true.

- there exist k pairwise disjoint elements of $\mathcal{S}$

- there exists a subset $Z \subseteq V(G)$ $|Z| \leqslant (k-1)(w+1)$ and $Z \cap S \neq \emptyset$ for each $S \in \mathcal{S}$.

Hint: We may as well assume that G is a w - tree. First consider the case $w = 1$ — that is — G is a tree. The Erdős - Pósa property says the following. Let $\mathcal{A}$ be a collection of subtrees of G. For every k either $\mathcal{A}$ has k elements that are vertex - disjoint or G has a subset of less than k vertices which hits every element of $\mathcal{A}$.

The subtrees are vertices in a chordal graph. When every pair of subtrees intersects then they have a point in common.

Let's get to the point.

A graph is biconnected if it has no separator with less than two vertices — that is — the graph is connected and has no cutvertex.

Lemma 4.179. *There exists a function* $f : \mathbb{N} \to \mathbb{N}$ *with the following property. Let* D *be a digraph whose underlying graph is biconnected and assume that* D *has no* $(t + 1)$ - *alternating path. There exists a set* $Z \subseteq V(D)$ $|Z| \leqslant f(t)$ *such that every* t - *alternating path* P *satisfies one of the following two statements.*

- *there is a series parallel separation* (A, B) *with* $P \subseteq A$

- $V(P) \cap Z \neq \emptyset$.

Proof. If a digraph has two vertex disjoint threads and $2t+3$ vertex disjoint threads that run between them then D has a $(t+1)$ - alternating path. This implies that the underlying graph of D has no subdivision of a $2 \times k$ -wall (for sufficiently large k). By the grid minor - theorem there exists $w \in \mathbb{N}$ such that D has treewidth w.

Let $f(t) = 4(w+1)$. (This is a function of t; we show below that this works.)

Let P be a t - alternating path in D.

- when t is odd then let m denote the pivot in the middle

- when t is even then let m and m' denote the two middle - pivots. By symmetry we may assume that m is a sink and m' is a source.

Let P_1 and P_2 be two vertex - disjoint t - alternating paths in D. Denote the pivots in the middle of P_i as m_i and m'_i ($i \in \{1,2\}$).

Let P be a thread that that runs between P_1 and P_2.

1. if t is odd then $V(P_1) \cap V(P) = m_1$ and $V(P_2) \cap V(P)$ is between the $(\lceil \frac{t}{2} \rceil - 1)^{\text{th}}$ pivot and the $(\lceil \frac{t}{2} \rceil + 1)^{\text{th}}$ pivot of P_2 or vice versa. Furthermore, if $V(P) \cap P_2 \neq \{m_2\}$ then P is a directed path

2. if t is even then P is a directed path between m_1 and m'_2 or vice versa.

Exercise: Let D be an orientation of a ladder, say with $2t+3$ steps. Show that D has an $(t+1)$ - alternating path.

Hint: There are $t+2$ steps whose orientation from one stringer to the other is the same.

Let $k \in \mathbb{N}$ and let G be a graph. Either the treewidth of G is at most k or G has an $f(k) \times f(k)$ - grid as a minor, for some function $f : \mathbb{N} \to \mathbb{N}$. This is the <u>grid minor - theorem</u>.

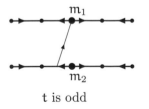

t is odd

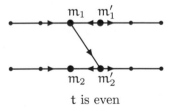

t is even

Figure 4.18: The figure illustrates two disjoint t - alternating paths — P_1 and P_2 — interacting with a thread P that runs between them.

By assumption the underlying graph of D is biconnected and since there are no three disjoint threads between P_1 and P_2 it follows that there exists a separation (A, B) of order two with $P_1 \subseteq A$ and $P_2 \subseteq B$. — Furthermore — there exist two disjoint directed paths Q_1 and Q_2 that run between P_1 and P_2 and which satisfy the following.

- if t is even then Q_1 is a directed path from m_1 to m_2' and Q_2 is a directed path from m_2 to m_1'

- if t is odd then Q_1 has endpoint m_1 and Q_2 has endpoint m_2. Furthermore, the other end of Q_1 is not m_2 and the other end of Q_2 is not m_1.

We show the following.

For any FIVE vertex - disjoint t - alternating paths $P_1, \cdots, P_5$ there exists a series parallel separation (A, B) with $P_i \subseteq A$ for some $i \in [5]$.

Assume the paths exist. Let (A, B) be a separation of order two —say $V(A) \cap V(B) = \{s, t\}$ — such that $P_1 \subseteq A$ and $P_2 \subseteq B$. By assumption (A, B) and (B, A) are not series parallel separations. At most two of the three other paths can intersect $\{s, t\}$. Assume $V(P_3) \cap \{s, t\} = \varnothing$ and $P_3 \subseteq A - s - t$.

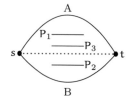

Let $Q_{3,1}$ be a directed path from P_3 to P_1 as mentioned above. Then $Q_{3,1} \subseteq A$.

To see that first assume that $Q_{3,1}$ contains $\{s, t\}$. Then we can replace the part that passes through s and t by a thread in B. The result should be a directed path from m_3 to $P_1 \setminus \{m_1\}$. However, (B, A) is not a series parallel separation and so B contains a thread between s and t that is not a directed path. This proves $Q_{3,1} \subseteq A$.

We claim that there is a separation (A', B') of order two with $V(A' \cap B') = \{m_1, m_2\}$ and

$$P_1 \cup P_3 \quad \subseteq \quad A' \quad \text{and} \quad P_2 \quad \subseteq \quad B'$$

To prove that we show that there is no thread in $D - \{m_1, m_2\}$ between m_3 and $P_2 \setminus \{m_2\}$. That is so because a merge of such a thread with $Q_{3,1}$ would be a thread between $P_1 \setminus \{m_1\}$ and $P_2 \setminus \{m_2\}$ which is a contradiction. So no component of $D - \{s, t\}$ intersects

P_3 and $P_2 - m_2$ and no component of $D - \{s, t\}$ intersects $P_1 - m_1$ and $P_2 - m_2$. This proves the claim.

If $Q_{2,3} \cap P_1 = \varnothing$ then $Q_{2,3} \subseteq A'$. Merge $Q_{2,3}$ with a thread in B' to obtain thread $m_2 \rightsquigarrow m_1$. Since (B', m_1, m_2) is not a one way series parallel triple there is a thread $m_2 \rightsquigarrow m_1$ which is not a directed path. So $Q_{2,3} \cap P_1 \neq \varnothing$.

If $Q_{2,3} \cap P_1 \neq \varnothing$ then let P'' be the subthread of $Q_{2,3}$ from $P_3 - m_3$ to P_1. Then P'' has end m_1 and $P'' \subseteq A'$ (since $m_2 \notin V(P'')$). The concatenation $P'' \cup Q_{1,2}$ is a thread from $P_3 - m_3$ to $P_2 - m_2$ and this is a contradiction.

Let $\mathcal{S}$ be the collection of vertex - sets of t - alternating paths P for which there is no series parallel separation (A, B) with $P \subseteq A$. By Exercise 4.105 there exists a set Z of vertices in D with $|Z| \leqslant 4(w+1)$ which hits every set $S \in \mathcal{S}$.

This proves the lemma. $\qquad\square$

AHEAD LIES A CLEAR ROAD TO GLORY; we should examine the extreme series parallel separations.

Definition 4.180. A series parallel separation (A, B) of a digraph is maximal if there exists no series parallel separation (A', B') in the digraph with $A \subset A'$.

Lemma 4.181. *Let D be a digraph whose underlying graph is biconnected. Assume that $D \neq X \cup Y$ for one way series parallel triples (X, s, t) and (Y, t, s). If (A_i, B_i) are two distinct maximal series parallel separations then*

$$A_1 \ \subseteq \ B_2 \quad and \quad A_2 \ \subseteq \ B_1.$$

Proof. For $i \in [2]$ let $A_i \cap B_i = \{s_i, t_i\}$ such that every thread $s_i \rightsquigarrow t_i$ in A_i is a directed path from s_i to t_i.

Assume that $t_2 \in V(A_1)$ and that $s_2 \in V(B_1)$ (see Figure 4.19). Let P_2 and P_2' be threads that run between s_2 and t_2 in A_2 and

B_2. Every thread in A_1 from s_1 to t_1 contains t_2 (since it connects $A_2 \setminus B_2$ with $B_2 \setminus A_2$) and $s_1 \in V(P_2)$.

It follows that $(A_1 \cup A_2, s_2, t_1)$ is a one way series parallel triple. By assumption $D \neq A_1 \cup A_2$ so $E(B_1 \cap B_2) \neq \varnothing$. This shows that $(A_1 \cup A_2, B_1 \cap B_2)$ is a series parallel separation and $A_1 \subset A_1 \cup A_2$ (since $s_2 \notin V(A_1)$). This contradicts the assumption that (A_1, B_1) is maximal.

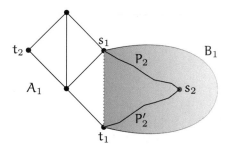

Figure 4.19: Illustration of Case 1

<u>Case 2</u>: Assume that $\{s_2, t_2\} \subseteq V(A_1)$. Then $B_1 \subseteq A_2$ or $B_1 \subseteq B_2$. When $B_1 \subseteq B_2$ then $A_2 \subseteq A_1$ and this contradicts that (A_2, B_2) is maximal.

So we have $B_1 \subseteq A_2$. Then (B_1, t_1, s_1) is a one way series parallel triple and so D is the union of one way series parallel triples A_1 and B_1. This is a contradiction.

Since (A_1, B_1) is maximal $A_1 \not\subseteq A_2$ and so $A_1 \subseteq B_2$.

This proves the lemma. $\qquad\qquad\qquad\qquad\qquad\qquad\qquad\qquad$ $\square$

TO SUMMARIZE: Let D be a digraph whose underlying graph is biconnected and assume that D is not a union of one way series parallel triples (X, s, t) and (Y, t, s). The collection $\mathcal{S}$ of maximal series parallel separations of D satisfies the following.

- for every series parallel separation (A, B) of D there exists $(A', B') \in \mathcal{S}$ with $A \subseteq A'$

- when $(A_1, B_1) \in \mathcal{S}$ and $(A_2, B_2) \in \mathcal{S}$ then $A_1 \subseteq B_2$ and $A_2 \subseteq B_1$.

4.15.7 Coda

In this section we prove Theorem 4.166.

IN CASE YOU LOST TRACK; it's the theorem below.

Theorem 4.182 (Liu and Muzi's theorem). *Let* $k \in \mathbb{N}$ *and let* (D_i) *be a sequence of digraphs without* k *- alternating path. Let* $(Q, \leqslant)$ *be a well quasi - order and for* $i \in \mathbb{N}$ *let* $\phi_i : V(D_i) \to Q$. *Then there exist* $j < j'$ *and a strong immersion* $\eta : D_j \to D_{j'}$ *such that for all* $x \in V(D_j)$

$$\phi_j(x) \quad \leqslant \quad \phi_{j'}(\eta(x)).$$

LET'S GET IN THE MOOD and start with an easy exercise.

"In the mood" is a tune by Glen Miller. In a future edition of this book we will let you listen to it!

Exercise 4.106

Let D be a digraph whose underlying graph is biconnected. Let r, x and y be three vertices of D and assume that every thread from r to $\{x, y\}$ is a directed path. Then for one of x and y there are directed paths to and from r.

HINT: See the figure.

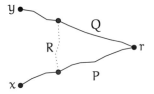

Figure 4.20: Let P and Q be threads from r to x and y with $V(P) \cap V(Q) = \{r\}$. (Exercise: show that P and Q exist.) Let R be a thread that connects $P \backslash r$ with $Q \backslash r$. (R exists.) When all threads from r to $\{x, y\}$ are directed paths then one endpoint of R must be one of x or y.

To prove Theorem 4.166 we order the set of rooted digraphs.

For $t, k \in \mathbb{N} \cup \{0\}$ let $\mathcal{F}_{t,k}$ be the set of those rooted digraphs (D, r) that satisfy the following properties.

here we go again ... see Lemma 4.176 on Page 268.

- the underlying graph of D is connected

- r is not a cutvertex

- D has no $(t+1)$ - alternating path

- no block of D has a t - alternating path

- no k - alternating path in D has r as an endpoint.

Define the classes $\mathcal{F}_t$, $\mathcal{F}_t^b$ and $\mathcal{F}_t^*$ as follows.

- $\mathcal{F}_t$ is the set of rooted digraphs of which the underlying graph is connected and which has no t - alternating path

- $\mathcal{F}_t^b$ is the set of rooted digraphs with no t - alternating path and of which the underlying graph is biconnected. [126]

- $\mathcal{F}_t^*$ is the set of rooted digraphs without t - alternating path.

[126] This includes the case where the underlying graph is one vertex or two vertices that are adjacent: a graph is biconnected if it is connected and has no cutvertex.

Exercise 4.107

1. $\mathcal{F}_t^b \subseteq \mathcal{F}_t \subseteq \mathcal{F}^*$

2. $\varnothing \subseteq \mathcal{F}_{t,0} \subseteq \cdots \subseteq \mathcal{F}_{t,t+1} = \sum_{k \geqslant 0} \mathcal{F}_{t,k}$.

Exercise 4.108

If $\mathcal{F}_{t,t+1}$ behaves well then $\mathcal{F}_t \cup \mathcal{F}_t^*$ behaves well.

HINT: By Higman's lemma if $\mathcal{F}_t$ behaves well then so does $\mathcal{F}_t^*$. So it is suffient to prove that $\mathcal{F}_t$ behaves well. To show this apply Higman's lemma (on words that are composed of letters in the well - behaved set $\mathcal{F}_{t,t+1}$).

Lemma 4.183. *If $\mathcal{F}_t^b$ behaves well then so does $\mathcal{F}_{t,k}$ for every integer $k \geqslant 0$.*

Proof. We leave it as an exercise to check that $\mathcal{F}_{t,0}$ behaves well.

HINT: $\mathcal{F}_{t,0} = \varnothing$.

We proceed by induction on k and assume that $\mathcal{F}_{t,k-1}$ behaves well.

Let $(Q, \leqslant)$ be a well quasi - order. Let (D_i, r_i) be a sequence of rooted digraphs in $\mathcal{F}_{t,k}$ and let $\phi_i : V(D_i) \to Q$.

Let S_i be a minimal set of cutvertices x of D_i that root some branch $(B, x) \in \mathcal{F}_t^b \cup \mathcal{F}_{t,k-1}$ and that covers all branches of D_i that are in $\mathcal{F}_{t,k-1}$ — that is — for every branch (B, x) of D_i that is in $\mathcal{F}_{t,k-1}$ there is a $(B', x') \in \mathcal{F}_t^b \cup \mathcal{F}_{t,k-1}$ with $B \subseteq B'$ and $x' \in S_i$.

By the induction assumption and by Higman's lemma the branches of D_i that are in $\mathcal{F}_t^b \cup \mathcal{F}_{t,k-1}$ are well - behaved.

Let D_i' be the digraph obtained from D_i by removing the internal vertices of branches at vertices $x \in S_i$ that are in $\mathcal{F}_t^b \cup \mathcal{F}_{t,k-1}$.

Label the vertices of D_i' with elements of a well quasi - order $(Q', \leqslant')$ as follows.

1. if $x \notin S_i$ then label x with $\phi_i(x)$

2. if $x \in S_i$ then label x with a pair $(\phi_i(x), \phi_i(B))$ where B is the union of branches at x that are in $\mathcal{F}_t^b \cup \mathcal{F}_{t,k-1}$.

Quasi - order pairs by the Cartesian product of the components.

If D_i' is biconnected then it is in $\mathcal{F}_t^b$ (since it is in $\mathcal{F}_{t,k}$). So — since $\mathcal{F}_t^b$ behaves well — if there are an infinite number of D_i' that are biconnected then we are done. — Henceforth — we assume that all elements of the sequence (D_i') have cutvertices.

The following claim is easily checked. When x is a cutvertex of D_i then all threads that run between r_i and x are directed paths and they all run in the same direction.

HINT: Let x be a cutvertex such that some thread $r_i \rightsquigarrow x$ is not directed. By definition of S_i $(B, x) \notin \mathcal{F}_{t,k}$ so there is a $(k-1)$ - alternating path in B that ends in x. Then there is a k - alternating path that ends in r_i. This contradicts that $(D_i, r_i) \in \mathcal{F}_{t,k}$.

Every block of D_i' has at most two cutvertices and the block that contains r_i contains at most one cutvertex of D_i'. To see that use Exercise 4.20.

It follows that (D_i', r_i) is an $\mathcal{F}_t^b$ - series parallel tree. We show that the set of $\mathcal{F}_t^b$ - series parallel trees behaves well. Let $\mathcal{F}'$ be the set of one way series parallel - triples (B, x, y) with $(B, x) \in \mathcal{F}_t^b$ and $y \in V(B) \setminus x$. These series parallel triples are in A_t. The set of all the truncations of the element of $\mathcal{F}'$ are in A_t (see Lemma 4.176). Thus $\mathcal{F}'$ and all truncations are well‑similated. By Lemma 4.175 this proves the claim. □

Exercise 4.109

Let $\mathcal{F}$ be a collection of rooted digraphs and assume that $\mathcal{F}$ behaves well. For $s \in \mathbb{N}$ let $\mathcal{F}^s$ be the collection of rooted digraphs (D, r) for which there exists

$$X \subseteq V(D) \quad |X| \leqslant s \quad r \in X \quad \text{and} \quad (D - X, r') \in \mathcal{F}$$

$$\text{for some } r' \in D \setminus X. \quad (4.21)$$

Show that $\mathcal{F}^s$ behaves well.

HINT: Let $(Q, \leqslant)$ be a well quasi‑order and let $((D_i, r_i))$ be a sequence in $\mathcal{F}^s$ and write

$$X_i \quad = \quad \{u_{i,1}, u_{i,2}, \cdots, u_{i,s}\} \quad \text{where } u_{i,1} = r_i.$$

For $x \in V(D_i - X_i)$ define

$$\phi_i'(x) \quad = \quad (\phi_i(x), a_1, b_1, \cdots, a_s, b_s),$$

where a_ℓ is the number of edges $u_{i,\ell} \to x$ and b_ℓ is the number of edges $x \to u_{i,\ell}$. Define a useful well quasi‑order to label the vertices of $D_i - X$.

Lemma 4.184. $\mathcal{F}_t^b$ *behaves well for all* $t \in \mathbb{N}$.

Proof. We prove this by induction on t. For $t = 1$ the claim is proved in Exercise 4.101 on Page 262.

Assume that $\mathcal{F}_{t-1}^b$ behaves well. By Exercise 4.183 and Lemma 4.108 $\mathcal{F}_{t-1}^*$ behaves well.

Let $(Q, \leqslant)$ be a well quasi‑order; let $((D_i, r_i))$ be a sequence in $\mathcal{F}_t^b$ and let $\phi_i : V(D_i) \to Q$. We show that there exist $j < j'$ and a strong immersion $\eta : (D_j, r_j) \to (D_{j'}, r_{j'})$ which satisfies $\phi_j(x) \leqslant \phi_{j'}(\eta(x))$ for all $x \in V(D_j)$.

Recall: $\mathcal{F}_t^b$ is the collection of rooted digraphs (D, r) that have no t‑alternating thread and of which the underlying graph is biconnected. For $t = 1$ these are obtained from a directed path or cycle by multiplication of edges.

Assume that for infinitely many i $D_i = X_i \cup Y_i$ for one way series parallel triples (X_i, s_i, t_i) and (Y_i, t_i, s_i). Then we are done by Lemma 4.176. So — by the summary on Page 274 — we may assume that every D_i has a collection of separations S_i that satisfy

Remove any D_i that is a union of one way series parallel triples. We may assume that this removes only a finite number of elements from the sequence. So we are left with an infinite sequence; which we simply call (D_i).

- $(A, B) \in S_i$ is a series parallel separation of D_i

- if $(A_1, B_1) \in S_i$ and $(A_2, B_2) \in S_i$ then $A_1 \subseteq B_2$ and $A_2 \subseteq B_1$

- if (A, B) is a series parallel separation of D_i then there exists $(A', B') \in S_i$ with $A \subseteq A'$.

S_i is the set of maximal series parallel separations of D_i.

> By Lemma 4.179 there exist $N \in \mathbb{N}$ and $Z_i \subseteq V(D_i)$ with $|Z_i| \leqslant N$ which hits every $(t-1)$ - alternating path P for which there is no series parallel separation (A, B) with $P \subseteq A$.

Let's get started.

Let $(A, B) \in S_i$. Replace A with a <u>handle</u> which is a directed P_5 that runs between the two terminals of A and that has a multiplicity on its edges. The multiplicity of the end - edges are the degrees in A of the two terminals. The multiplicity of the two middle edges is the number of edge - disjoint directed paths in A that run between the two terminals.

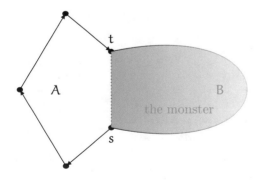

Figure 4.21: Replacement of A with a handle. Multiplicities are not shown.

Let D_i' be obtained from D_i by replacing each A_i of the separations $(A_i, B_i) \in S_i$ that satisfies $|V(A_i) \setminus V(B_i)| \geqslant 2$ with a handle.

WE NEED TO SUPPLY D_i' WITH A ROOT. If $r_i \in B_i$ then let $r_i' = r_i$ and if $r_i \in A_i \setminus B_i$ then let r_i' be an arbitrary vertex of D_i'.

Let Z_i' be the following set of vertices in D_i'.

1. the vertices of Z_i that are in D_i'

2. r_i'

3. the two terminals s and t for every one way series parallel triple (A, s, t) for which

   - $A \cup B = D_i$

   - $(A, B) \in S_i$

   - $V(A \cap B) = \{s, t\}$

   - $|V(A) \setminus V(B)| \geqslant 2$

   - Z_i has a vertex in $A \setminus B$.

... for which A has been replaced by a handle and for which $Z_i \cap (A \setminus B) \neq \varnothing$.

Then $Z_i' \subseteq V(D_i) \cap V(D_i')$ and $|Z_i'| \leqslant 2 \cdot |Z_i| + 1 \leqslant 2 \cdot N + 1$.

NOTICE THAT Z_i' hits every $(t-1)$ - alternating path in D_i'.

Let $D_i^* = D_i' - Z_i'$ and let r_i^* be an arbitrary vertex of D_i^*. By the previous observation $(D_i^*, r_i^*) \in \mathcal{F}_{t-1}^*$.

Let $\mathcal{F}^*$ be the set of rooted digraphs (D, r) which satisfy

EXERCISE! Hint: Assume there is a $(t-1)$ - alternating path P' in D_i' that misses Z_i'. Replace parts in handles by threads in D_i and construct an $(t-1)$ - alternating path in D_i that misses Z_i.

1. D has a set Z of vertices with $r \in Z$ and $|Z| \leqslant 2 \cdot N + 1$

2. $(D - Z, r') \in \mathcal{F}_{t-1}^*$ for some $r' \in V(D) \setminus Z$.

Then $\mathcal{F}^*$ behaves well and $(D_i', r_i') \in \mathcal{F}^*$.

Exercise 4.109 shows that $\mathcal{F}^*$ behaves well.

TO EXPLOIT THE FACT that $\mathcal{F}^*$ behaves well the vertices of (D_i', r_i') are now labeled with elements of a well quasi - order — say — $\phi_i' : V(D_i') \to Q'$. First we supply Q' with an element that is incomparable to all others. This element of Q' is used to label r_i'. For vertices x of D_i' that are in D_i define $\phi_i'(x) = \phi_i(x)$.

It remains to label the vertices of handles that replace one way series parallel triples; say (A, s, t) in D_i. The midpoint of such a handle is labeled as $((A, s, t), \phi_i)$.

Fix a partition $\{S, T\}$ of $V(A)$ with $s \in S$ and $t \in T$ and such that the crossing edges form a minimum cut. Label the two neighbors of the midpoint in the handle with the two ~~truncations~~ of which the vertices are labeled by ϕ_i.

$\mathcal{F}^*$ behaves well; (D'_i) is a sequence in $\mathcal{F}^*$ and $\phi'_i : V(D'_i) \to Q'$ for a well quasi - order $(Q', \leqslant')$. — Thus — there exist $j < j'$ and a strong immersion $\eta' . V(D'_j) \to V(D'_{j'})$ which respects $(Q', \leqslant')$. It follows easily that there is a strong immersion $\eta : (D_j, r_j) \to (D_{j'}, r_{j'})$ which respects $(Q, \leqslant)$.

This proves the lemma.   □

Lemma 4.176 shows that one way series parallel triples without k - alternating threads and their trunca- tions are well - simulated.

HOORAY! We're done.

Hip, hip!

Theorem 4.185. *Let* $k \in \mathbb{N}$ *and let* (D_i) *be a sequence of digraphs without* k - *alternating path. Let* $(Q, \leqslant)$ *be a well quasi - order and for* $i \in \mathbb{N}$ *let* $\phi_i : V(D_i) \to Q$. *Then there exist* $j < j'$ *and a strong immersion* $\eta : D_j \to D_{j'}$ *such that for all* $x \in V(D_j)$

$$\phi_j(x) \quad \leqslant \quad \phi_{j'}(\eta(x)).$$

Proof. By Lemma 4.184 $\mathcal{F}^b_t$ behaves well and so $\mathcal{F}^*_t$ behaves well (by Exercise 4.108 and Lemma 4.183).

This proves the theorem.   □

4.15.8 Exercise

A permutation graph is an intersection graph of a set of straight line - segments with their endpoints on two parallel lines.

Figure 4.22: The figure shows a permutation dia- gram. Crossing line seg- ments represent adjacent vertices in the permutation graph.

A graph G is a permutation graph if and only if G and $\bar{G}$ are comparability graphs — so — a permutation graph can be represented as a tournament with a 2-coloring of its edges such that every color is transitive.

Exercise 4.110

Let $(P_i)_{i \in \mathbb{N}}$ be a sequence of permutation graphs. Show that there exist $j < j'$ such that P_j immerses strongly in $P_{j'}$.

Show that the class of AT - free graphs is well quasi - ordered by strong immersions.

> A graph is AT - free if it has no asteroidal triple — that is — if it has no three vertices of which every pair is connected by a path that avoids the closed neighborhood of the third.

4.16 Asteroidal sets

Definition 4.186. Let G be a graph. A set $A \subseteq V$ is an asteroidal set if for each vertex $a \in A$ the set $A \setminus \{a\}$ is contained in a component of $G - N[a]$.

ASTEROIDAL SETS WITH 3 VERTICES are called asteroidal triples. — For example — consider a claw and subdivide every edge one time. The set of leaves of this tree is an asteroidal triple. Another example is an independent set in C_6 (or the simplicials in a 3-sun). Gallai presents a list of the minimal graphs that have an asteroidal triple.

The concept was used by Lekkerkerker and Boland to characterize interval graph in the following manner.

> A graph is an interval graph if and only if it is chordal and has no asteroidal triples.

4.16.1 AT - free graphs

In this section we have a look at the structure of graphs that do not have an asteroidal triple. CLEARLY (by the characterization of Lekkerkerker and Boland) interval graphs are graphs without asteroidal triple. Another example of a class of graphs that are AT-free is the class of permutation graphs.

All complements of comparability graphs are AT-free. To see that, use the fact that cocomparability graphs are intersection graphs of continuous functions $f : [0, 1] \to \mathbb{R}$. When 3 functions pairwise don't intersect then one is between the other two — and so — its closed neighborhood hits every path that runs between the outer two.

4.16.2 Independent set in AT-free graphs

COMPUTING ω IS NP-COMPLETE ON AT-FREE GRAPHS. That is so because α is NP-complete on triangle-free graphs. In this section we show that there is a polynomial algorithm to compute the independence number α on AT-free graphs.

The algorithm computes (recursively) the following numbers.

Exercise 4.111

Let G be a graph. Then

$$\alpha(G) \;=\; 1 + \max_{x \in V} \sum_i \alpha(C_i)$$

where the C_i are the components of $G - N[x]$.

Let G be AT-free and let x and y be nonadjacent vertices in G. The <u>interval</u> $I(x, y)$ is the set of all vertices that are between x and y.

Exercise 4.112

Let G be AT-free. Let $x \in V$ and let C be a component of $G - N[x]$.

$$\alpha(C) \quad = \quad 1 + \max_{y \in C} \left\{ \alpha(I(x,y)) + \sum_i \alpha(C_i) \right\}$$

where the C_i are the components of $G - N[y]$ that are

contained in C.

THE FINAL STEP IS TO DECOMPOSE THE INTERVALS.

Exercise 4.113

Let G be AT-free and let $I(x,y)$ be an interval in G. When $I = \varnothing$ then $\alpha(I) = 0$. Otherwise

$$\alpha(I) \quad = \quad 1 + \max_{s \in I} \alpha(\, I(x,s)\,) + \alpha(\, I(y,s)\,) + \sum_i \alpha(C_i)$$

where $C_1, \cdots$ are the components of $G - N[s]$ that are

contained in I.

Exercise 4.114

Prove the following theorem.

Theorem 4.187. *There exists an* $O(n^4)$ *algorithm to compute the independence number in* AT*-free graphs.*

Exercise 4.115

Show that the computation of the clique number ω is NP - complete on AT - free graphs.

Hint: The independence number α is NP - complete for the class of triangle - free graphs.

4.16.3 Exercise

AT - FREE GRAPHS ARE χ - BOUNDED — that is — there exists
a function $f : \mathbb{N} \to \mathbb{N}$ which satisfies

$$\chi \;\;\leqslant\;\; f(\omega) \qquad \text{for all } \textsf{AT} \text{ - free graphs.}$$

That is so because AT - free graphs do not contain a subdivision
of a claw as an induced subgraph. Kierstead and Penrice showed
(in 1994) that the class of graphs without subdivision of a claw is
χ - bounded.

Remark 4.188. The Gyárás - Sumner conjecture suggests that for
every tree T the class of graphs that do no contain T as an induced
subgraph is χ - bounded.

Remark 4.189. WE ARE NOT AWARE of any hereditary class of
graphs which is χ - bounded but not polynomially so.

Exercise 4.116

Let $k \in \mathbb{N}$. Show that there is a polynomial - time algorithm to
check if $\chi \leqslant k$ for AT - free graphs.

A CONFLICT - FREE COLORING of a graph G is a coloring of
its vertices such that every closed neighborhood has a uniquely
colored vertex. Let $\kappa(G)$ denote the mimimal number of colors
needed in a conflict - free coloring of G.

Exercise 4.117

Show that $\textsf{t}\textsf{K}_2$ - free graphs satisfy $\kappa \leqslant 2 \cdot \textsf{t} - 1$.

Exercise 4.118

Show that circle graphs satisfy

$$\kappa \;\;\leqslant\;\; 28 \cdot \omega.$$

HINT: First show that permutation graphs satisfy $\kappa \leqslant 4$. To see that make use of the fact that permutation graphs have a shortest path that is dominating. Davies and McCarty show that the vertex set of a circle graph can be partitioned into 7ω parts that induce permutation graphs.

Exercise 4.119

PROVE OR DISPROVE: there exists $k \in \mathbb{N}$ such that every circle graph can be colored conflict free with k colors.

Remark 4.190. For any graph H the class of graphs that do not contain H as a vertex minor has bounded rankwidth if and only if H is a circle graph.

J. Geelen, O. Kwon, R. McCarty, and P. Wollan, The grid theorem for vertex - minors. Manuscript on arXiv: 1909.08113, 2020.

4.16.4   Bandwidth of AT-free graphs

Definition 4.191. A layout of a graph G is an ordering of its vertices $L : V \leftrightarrow [n]$. The width of L is 0 if $E = \varnothing$ and otherwise it is

$$\max \{ |L(x) - L(y)| \mid \{x, y\} \in E \}.$$

The bandwidth of G is the minimal width of a layout of G. [127]

In this section we prove the following theorem.

Theorem 4.192. *There exists a linear - time algorithm to approximate the bandwidth of AT-free graphs with worst - case performance ratio 6.*

---

[127] When a graph has small bandwidth then the rows and columns of its adjacency matrix can be permuted so that all 1s appear in a narrow band around the diagonal.

To prove the performance ratio we need a lower bound.

Lemma 4.193. *Let* G *be a graph. Let* $\{x, y\} \in E$. *Then*

$$\mathsf{bw}(G) \quad \geqslant \quad \frac{1}{3} \cdot (\, |N(x) \cup N(y)| - 1 \,).$$

Proof. Let L be an optimal layout. Assume $L(x) < L(y)$. Consider the three sets S_1, S_2, and S_3:

$$
\begin{aligned}
S_1 &= \{z \in N \mid L(z) \leqslant L(x)\} \\
S_2 &= \{z \in N \mid L(x) \leqslant L(z) \leqslant L(y)\} \\
S_3 &= \{z \in N \mid L(z) \geqslant L(y)\}
\end{aligned}
$$

where $N = N(x) \cup N(y)$.

Then

$$\mathsf{bw}(G) \geqslant \max_i |S_i| - 1 \quad \text{and} \quad \sum_i |S_i| = |N| + 2.$$

There must exist i such that $|S_i| - 1 \geqslant \frac{1}{3}(|N| - 1)$.

This proves the lemma. $\qquad\square$

Definition 4.194. A caterpillar is a tree with a dominating path.

The vertices of the caterpillar that are not in the dominating path are called the feet of the caterpillar. [128]

Exercise 4.120

A tree is a caterpillar if and only if it does not contain the tree obtained from a claw by subdividing each edge one time.
Show that a tree is AT-free if and only if it is a caterpillar.

Lemma 4.195. *There exists an* $O(n)$ *algorithm to approximate the bandwidth of caterpillars within a factor* $3/2$.

Proof. Let $[b_1 \cdots b_\ell]$ be the dominating path of a caterpillar T and let d_i be the number of feet attached to b_i. Define

$$L(b_i) \quad = \quad \left\lfloor \frac{d_i}{2} \right\rfloor + \sum_{j<i} d_j.$$

For feet z adjacent to b_i let

$$L(z) \quad \in \quad \left\{ L(b_i) - \left\lfloor \frac{d_i}{2} \right\rfloor, \quad \cdots \quad , L(b_i) + \left\lceil \frac{d_i}{2} \right\rceil \right\}.$$

The width of L is

$$\max_i L(b_{i+1}) - L(b_i) \quad = \quad \max_i \left\lceil \frac{d_i}{2} \right\rceil + \left\lfloor \frac{d_{i+1}}{2} \right\rfloor.$$

By Lemma 4.193

$$bw(T) \quad \geqslant \quad \frac{1}{3} \cdot \max_i d_i + d_{i+1} + 1.$$

This implies that the width of L is at most $\frac{3}{2} \cdot bw(T)$.

This proves the lemma. $\qquad\qquad\qquad\qquad\qquad\qquad\qquad\qquad \Box$

Exercise 4.121

Prove the following lemma.

Lemma 4.196. *Let G be a connected AT-free graph. There exists a spanning caterpillar T such that any adjacent pair in G is at distance at most 4 in T. This caterpillar can be found in linear time.* [129]

[129] The caterpillar T has $V(T) = V(G)$ (it spans V). The graph G is a (spanning) subgraph of T^4.

HINT: Use the fact that a connected AT-free graph has a dominating pair — that is — a pair of vertices such that any path that connects them is a dominating path.

WE ARE READY TO PROOF THEOREM 4.192.

Proof. Let G be AT-free. Let T be a spanning caterpillar such that adjacent vertices in G are at distance at most 4 in T. Let L be a layout of T of width at most $\frac{3}{2} \cdot bw(T)$.

Use L as a layout for G. We have

$$
\begin{aligned}
\mathsf{width}(G, L) \;&\leqslant\; \mathsf{width}(T^4, L) \\
&\leqslant\; 4 \cdot \mathsf{width}(T, L) \\
&\leqslant\; 4 \cdot \frac{3}{2}\mathsf{bw}(T) \\
&\leqslant\; 6 \cdot \mathsf{bw}(G).
\end{aligned}
$$

This proves the theorem. □

Remark 4.197. There exists an $O(m+n \log n)$ algorithm to compute the bandwidth on caterpillars. Alternatively there exists a $O(n^3)$ algorithm that approximates the bandwidth of AT-free graphs within a factor 2.

The bandwidth problem remains NP-complete on cobipartite graphs (which are AT-free). (For cobipartite graphs the bandwidth equals the treewidth of the graph.)

Another way to approximate the bandwidth of AT-free graphs is via the computation of a minimal triangulation. Let G be AT-free and let H be a minimal triangulation of G. Then the bandwidth of H is at most twice the bandwidth of G. To see that observe that AT-free graphs have no induced C_6. It follows that in any minimal separator S of G two nonadjacent vertices of S have a common neighbor in G. This shows that any two adjacent vertices of H that are not adjacent in G have a common neighbor in G. CONSEQUENTLY $\mathsf{bw}(H) \leqslant 2 \cdot \mathsf{bw}(G)$. — Finally — every minimal triangulation of G is an interval graph and there is an $O(n^2)$ algorithm to compute the bandwidth of interval graphs. [130]

[130] D. Kleitman and R. Vohra, *Computing the bandwidth of interval graphs*. SIAM Journal on Discrete Mathematics **3** (1990), pp. 373–375.

Exercise 4.122

A graph is AT-free if and only if every minimal triangulation is an interval graph.

4.16.5   Dominating pairs

A connected graph with at least two vertices and without asteroidal triples has a <u>dominating pair</u> — that is — a pair of vertices s and t with the property that every $s \rightsquigarrow t$ - path in the graph is a dominating set.

4.16.6   Antimatroids

Let V be the set of vertices of a graph G. A <u>betweenness</u> relation in G is a collection of rooted sets $\mathcal{K} = \{(K, r)\}$ where $K \subseteq V$ and $r \in K$. A betweenness relation $\mathcal{K}$ defines a convexity: a set $C \subseteq V$ is <u>convex</u> if

$$K \setminus r \subseteq C \quad \Rightarrow \quad r \in C$$

for every betweenness $(K, r) \in \mathcal{K}$.

Definition 4.198. Let V be a finite set and let $\mathcal{C}$ be a collection of subsets of V. The set system $(V, \mathcal{C})$ is a <u>convex geometry</u> if

1. $\varnothing \in \mathcal{C}$ and $V \in \mathcal{C}$

2. if $A \in \mathcal{C}$ and $B \in \mathcal{C}$ then $A \cap B \in \mathcal{C}$

3. if $A \in \mathcal{C}$ and $A \neq V$ then there exists $x \in V \setminus A$ such that $A \cup x \in \mathcal{C}$.

Chang et al. proved the following characterization of AT - free graphs.

Theorem 4.199. *There exists a betweenness relation such that the collection of convex sets in a graph is a convex geometry if and only if the graph has no asteroidal triple.*

The betweenness relation consists of rooted sets with three pairwise nonadjacent vertices for which there is a path from the root to each end that avoids the neighborhood of the other end.

When some vertex is between two others then it is one of the following.

1. the nose of a bull

2. a root of a 6 - chain

3. a midpoint of P_5

4. a pendant, adjacent to the midpoint of P_5.

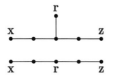 

Figure 4.23: The figure shows P_5, P_5 with a pendant, the bull and the 6-chain. The 'root' r is the element of the betweenness that is between the two 'ends' x and z.

An AT - free order is a shelling sequence of the convex geometry; it repeatedly removes vertices from the graph that are not between two others.

Algorithm 9: Compute an AT - free order

$\alpha \leftarrow \varnothing;$

while $\alpha \neq V$ **do**

    Choose $x \in V \setminus \alpha$ such that
      there is no betweenness (K, x) $K \subseteq V \setminus \alpha$ with root x;

    $\alpha \leftarrow \alpha x$

end while

Example 4.200. Consider a shelling of a poset which eliminates elements that have no descendants. The sequences are the words of an antimatroid. Poset antimatroids have a betweenness relation with only two elements — namely — the cover - relation of the poset.

If the poset is a rooted tree then the antimatroid is the collection of elimination orders which remove leaves until there are no more vertices. The betweenness relation is the parent relation of the tree.

Exercise 4.123

Let G be a chordal graph. The collection of simplicial elimination orders of G is an antimatroid. Please describe a concise betweenness relation that defines this antimatroid.

If a graph has no root of a P_5, bull or 6 - chain then it is AT - free and any order of the vertices is an AT - free order.

Lemma 4.201. *Let G be a graph and assume that G is prime with respect to modular decomposition. If G has no induced P_5, bull or 6 - chain then any independent set in G has at most two elements.*

Proof. Assume G is prime and has no induced P_5 or bull. Maffray shows that either G is the complement of a graph without triangles or G has no house (that is the complement of P_5) or C_5. [131]

[131] The refences are listed on Page 294.

Fouquet and Vanherpe show that if a graph is prime and has no C_5, P_5, house or bull then it is a chain graph or the complement of a chain graph. In our case the graph has no 6-chain. This leaves complements of chain graphs (which includes the 4-chain P_4). — In any case — the complements are graphs without triangle. $\square$

4.16.7   Totally balanced matrices

Let H be a hypergraph. Its incidence matrix is the 0/1 - matrix of which the rows are indexed by the vertices of H and the columns are indexed by the hyperedges of H. An entry (x, e) of this matrix is 1 if the vertex x is in the edge e.

Definition 4.202. A hypergraph is <u>totally balanced</u> if the incidence matrix does not contain a submatrix of size at least 3 with no identical columns and with each row sum and column sum equal to 2.

Lemma 4.203. *Let* G *be connected and* AT *- free. Let* $\{s, t\}$ *be a dominating pair and let* P *be a shortest* $s \rightsquigarrow t$ *- path in* G*. Let* H *be the hypergraph with vertex set* $V(G)$ *and the following edges. For each* P_3 *in* P *the union of the closed neighborhoods is an edge of* H*. Then* H *is totally balanced.*

Proof. It is sufficient to to show that the hyperedges are a path decomposition of G — in other words — the graph becomes an interval graph if we make clique of all hyperedges.

Both endpoints of an edge are in the closed neighborhood of a P_3 in P — otherwise there is a cycle of length at least 6.

Clearly each vertex of G is in a consecutive set of hyperedges. $\square$

Remark 4.204. Strongly chordal graphs are chordal graphs without a sun. They have a simple elimination order; that is a simplicial elimination order that avoids taking out the nose of a bull, or a midpoint of P_5, or a pendant to a midpoint of P_5.

A net is a graph that consists of a clique and an independent set both of size at least 3 and a perfect matching between them. A net is the smallest strongly chordal graph that has a nose of a bull between any two of its ends. (Three leaves in a net are an asteroidal triple.)

A vertex is simple if for any two vertices x and y in its closed neighborhood

$$N[x] \subseteq N[y] \quad \text{or}$$
$$N[y] \subseteq N[x].$$

Exercise 4.124

The simple elimination orders of an interval graph are the words of an antimatroid. Describe a betweenness relation that defines this antimatroid.

Exercise 4.125

A paired dominating set in a graph is a dominating set that has a perfect matching.

1. Show that every graph without isolated vertices has a paired dominating set.

2. Show that there is a greedy algorithm to compute a paired dominating set in AT - free graphs of smallest size.

HINT: Let P be a dominating shortest path. This defines a path -
decomposition as in Lemma 4.203. Prove that there is a minimum
paired dominating set with a perfect matching of which every edge
hits P. (See Figures 4.24, 4.25 and 4.26.)

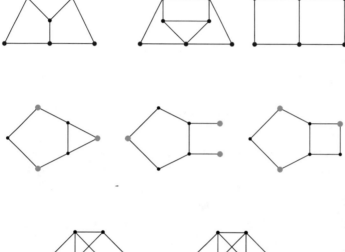

Figure 4.24: The figure
shows a betweenness involv-
ing a P_3 in P. (The red vertices are vertices of the shortest path P.)

Figure 4.25: The absence of
asteroidal triples limit the
neighborhood of a C_5

Figure 4.26: The figure
shows the connections be-
tween two P_4s.

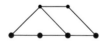  

Further reading

Alcón, L., B. Brešar, T. Gologranc, M. Gutierrez, T. Šumenjak, I. Pe-
terin, A. Tepeh, Toll convexity, *European Journal of Combinatorics*
46 (2015), pp. 161 – 175.

Beisegel, J., Characterising AT - free graphs with BFS. Manuscript on arXiv: 1807.05065, 2018.

Boyd, E. and U. Faigle, An algorithmic characterization of antimatroids, *Discrete Applied Mathematics* **28** (1990) pp. 197 – 205.

Chang, J., T. Kloks and H. Wang, Convex geometries on AT - free graphs and an application to generating the AT - free orders. Manuscript 2017.

Corneil, D. and J. Stacho, Vertex ordering characterizations of graphs of bounded asteroidal number, *Journal of Graph Theory* **78** (2015) pp. 61 – 79.

Farber, M., Domination, independent domination, and duality in strongly chordal graphs, *Discrete Applied Mathematics* **7** (1984) pp. 115 – 130.

Fouquet, J. and J. Vanherpe, Seidel complementation on $(P_5, \text{house}, \text{bull})$ - free graphs. Technical report Université d'Orléans, HAL - 00467642, 2010.

Korte, B., L. Lovász and R. Schrader, *Greedoids*, Springer, Series Algorithms and Combinatorics **4** 1980.

Lawler, E., Optimal sequencing of a single machine subject to precedence constraints, *Management Science* **19** (1973) pp. 544 – 546.

Maffray, F., Coloring (P_5, bull) - free graphs. Manuscript on arXiv: 1707.08918, 2017.

Nakamura, M., Excluded - minor characterizations of antimatroids arisen from posets and graph searches, *Discrete Applied Mathematics* **129** (2003) pp. 487 – 498.

Hoffman, A., A. Kolen and M. Sakarovitch, Totally balanced and greedy matrices. Technical report, Mathematical Centre, Amsterdam, 1980.

4.16.8   Triangle graphs

Circle graphs are the intersection graphs of chords of a circle.

Elmallah and Stewart introduced the class of k-polygon graphs. These graphs are the intersection graphs of chords in a k-sided polygon. Elmallah and Stewart show that k-polygon graphs can be recognized in polynomial time and that the domination problem can be solved in $O(n^{4k^2+3})$ time on k-polygon graphs.

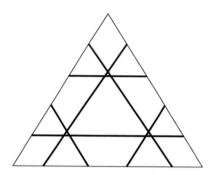

Figure 4.27: The figure shows some chords in a triangle. It is the model of a 3-sun.

Exercise 4.126

Define a betweenness which generates an antimatroid on triangle graphs. Design a greedy algorithm to compute γ on triangle graphs (γ is the domination number).

4.17 Sensitivity

| In 2019 Hao Huang proved the sensitivity conjecture! |

In this chapter we take a look at the proof.

A <u>decision tree</u> is an algorithm that evaluates a Boolean function $f : \{0,1\}^n \to \{0,1\}$ by a sequence of queries. A query reads one bit x_i of an input $x \in \{0,1\}^n$. The choice of a query depends on the outcome of previous queries.

Let T be a rooted binary, tree. The leaves are labeled as 0 or 1. Internal nodes (including the root) are labeled with variables x_i. Given input $x \in \{0,1\}^n$ the tree is evaluated as follows. If the root is a leaf then output its label 0 or 1. Otherwise query the value of the root variable x_i. If it is 0 then evaluate the left subtree — otherwise — evaluate the right subtree. The tree T 'computes' a Boolean function $f : \{0,1\}^n \to \{0,1\}$ if the algorithm described above gives output $f(x)$ for all $x \in \{0,1\}^n$. The depth of f is the smallest depth of a tree that evaluates f.

The depth of a decision tree is the largest number of queries made by the algorithm to evaluate $f(x)$ (over all $x \in \{0,1\}^n$). The <u>depth</u> $D(f)$ of a Boolean function f is the smallest depth over all decision trees that compute f.

Definition 4.205. Let $f \cdot \{0,1\}^n \rightarrow \{0,1\}$ be a Boolean function. The <u>sensitivity</u> of f at input $x \in \{0,1\}^n$ is the number of i's for which a flip of the ith element of x changes the value of $f(x)$. The <u>sensitivity</u> $s(f)$ of f is the largest sensitivity at input x over all $x \in \{0,1\}^n$.

Make sure you understand this definition properly: $\{0,1\}$ is an alphabet. Elements of $\{0,1\}^n$ are words of length n with letters in $\{0,1\}$. For $S \subseteq [n]$ let x^S be the word obtained from $x = x_1 \cdots x_n$ by <u>flipping</u> the value of x_i for $i \in S$. The sensitivity of f at x is the number of $i \in [n]$ for which $f(x) \neq f(x^{\{i\}})$.

In 2019 Hao Huang proved the following theorem.

Theorem 4.206 (The sensitivity theorem). *Let* $f : \{0,1\}^n \rightarrow \{0,1\}$ *be a Boolean function. Then* $s(f) \leqslant D(f)$ *and there exists a constant* c *such that*

$$D(f) = O(s(f)^c).$$

Exercise: Show that $D(f) \geqslant s(f)$.

4.17.1 What happened earlier …

For $x \in \{0,1\}^n$ and $S \subseteq [n]$ let x^S be the word obtained from x by flipping all bits in S.

Definition 4.207. The <u>block sensitivity</u> of a Boolean function f at $x \in \{0,1\}^n$ is the maximal number of disjoint subsets $B \subseteq [n]$ for which $f(x^B) \neq f(x)$. The block sensitity of f is the largest block sensitivity of $f(x)$ over all $x \in \{0,1\}^n$.

Noam Nisan showed (in 1989) the following sandwich

$$s(f) \leqslant bs(f) \leqslant D(f) = O(bs(f)^4).$$

Exercise: Show that $bs(f) \geqslant s(f)$.

Hao Huang proves the following theorem. — Notice — that this proves the sensitivity theorem.

Exercise: Show that this theorem proves the sensitivity theorem.

> **Theorem 4.208.** *For every Boolean function*
>
> $$s(f) \;\;\leqslant\;\; bs(f) \;\;\leqslant\;\; s(f)^4.$$

WE TAKE A LOOK AT THE PROOF — but first — let's do something else.

4.17.2   Cauchy's interlace lemma

Let A be a real symmetric $n \times n$ matrix. Then all eigenvalues are real numbers. A principle submatrix B is a submatrix of A on the same subset of rows and columns. Cauchy's interlace lemma says that the eigenvalues of A and B <u>interlace</u> which is defined as in the lemma.

Lemma 4.209. *Let* A *be a real symmetric* $n \times n$ *- matrix. Let* B *be a* $m \times m$ *principal submatrix of* A. *Let* $\lambda_1 \geqslant \cdots \geqslant \lambda_n$ *be the eigenvalues of* A *and let* $\mu_1 \geqslant \cdots \geqslant \mu_m$ *be the eigenvalues of* B. *Then for* $i \in [m]$:

$$\lambda_i \;\;\geqslant\;\; \mu_i \;\;\geqslant\;\; \lambda_{i+n-m}.$$

The eigenvalues of A and B are like shoelaces. Shoelaces 'interlace' (that is; they 'twine') to tie up your shoe.
When $m = n - 1$ then

$$\lambda_1 \geqslant \mu_1 \geqslant \lambda_2 \geqslant \cdots$$
$$\cdots \geqslant \mu_{n-1} \geqslant \lambda_n$$

4.17.3   Hypercubes

FOR A PROOF OF THE FOLLOWING LEMMA see eg the monograph by Brouwer and Haemers on spectra of graphs.

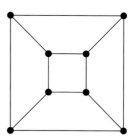

Figure 4.28: A hypercube

Lemma 4.210. *The spectrum of the hypercube* Q_n *consists of the numbers* $n - 2i$ *with multiplicity* $\binom{n}{i}$ *for* $i = 0 \cdots n$.

In his paper Huang Huo proves the following theorem.

Exercise: Prove Theorem 4.211 by using Lemma 4.210 and interlacing.

Hint: $\Delta(H) \geqslant \lambda_1(H)$.

Theorem 4.211 (The hypercube theorem). *Let* H *be an induced subgraph of the hypercube* Q_n *with* $2^{n-1} + 1$ *vertices. Then the largest degree in* H *satisfies*

$$\Delta(H) \;\geq\; \sqrt{n}$$

and — this inequality is tight when n *is a square.*

To prove Theorem 4.211 let's start with two exercises.

Exercise 4.127

Define a sequence (A_n) of $\{0, -1, +1\}$ - matrices as follows.

$$A_1 = \begin{pmatrix} 0 & 1 \\ 1 & 0 \end{pmatrix} \quad \text{and} \quad A_n = \begin{pmatrix} A_{n-1} & I \\ I & -A_{n-1} \end{pmatrix}.$$

If we flip all -1's in A_n to $+1$ we get the adjacency matrix of the hypercube Q_n.

Show that the eigenvalues of A_n are $\sqrt{n}$ and $-\sqrt{n}$ both of multiplicity 2^{n-1}.

Hint: A_n satisfies $A_n^2 = n \cdot I$.

Remark: An $n \times n$ conference matrix C has zeros on the diagonal and $+1$ or -1 everywhere else and satisfies $C^T C = (n-1)I$. Van Lint and Seidel show that a symmetric conference matrix can only exist if $n = 2 \bmod 4$ and $n - 1$ is a sum of two squares.

Exercise 4.128

Let G be a graph. Let A be a symmetric matrix whose rows and columns are indexed by $V(G)$ and which has entries in $\{0, -1, +1\}$ such that

$$A(x, y) \neq 0 \qquad \Rightarrow \qquad \{x, y\} \in E(G)$$

Then $\Delta(G) \geq \lambda_1(A)$. [132]

[132] $\lambda_1(A)$ is the largest eigenvalue of A and $\Delta(G)$ is the largest degree of a vertex in G.

The proof of the hypercube theorem

Remark: A matrix H is Hadamard if all its entries are $+1$ or -1 and $HH^T = n \cdot I$. When H is hadamard then so is $\begin{pmatrix} H & H \\ H & -H \end{pmatrix}$.

Proof. Let A_n be the $\{0, -1, +1\}$ - matrix as defined in Exercise 4.127. Change all the -1 - entries in A_n to $+1$. Then the matrix becomes the adjacency matrix of the hypercube Q_n. — So — we may assume that there is a 1-1 correspondence between rows and columns of A_n and Q_n.

Let H be an induced subgraph of Q_n with at least $2^{n-1}+1$ vertices. Let A_H be the principal submatrix of A_n whose rows and columns are indexed by the vertices of H. By Exercise 4.128

$$\Delta(H) \geqslant \lambda_1(A_H).$$

By Cauchy's interlace lemma:

$$\lambda_1(A_H) \geqslant \lambda_{2^{n-1}}(A_n) = \sqrt{n}. \qquad (4.22)$$

This proves the theorem. $\qquad\qquad\qquad\qquad\qquad\qquad\square$

It is easy to see that the inequality (4.22) is tight: Let H be the subgraph of Q_n induced by all vertices of even weight and one vertex of odd weight. Then H is a union of the star $K_{1,n}$ and isolated vertices. The largest eigenvalue is $\sqrt{n}$.

There are $\sum_i \binom{n}{2i} = 2^{n-1}$ vertices of even weight. They form an independent set in Q_n. Every vertex in Q_n of odd weight has a neighborhood of size n which is a set of vertices that all have even weight.

4.17.4   Möbius inversion

Let $f : \{0,1\}^n \to \mathbb{R}$ be a map. We show that f can be represented as a polynomial in n variables $x_1, \cdots, x_n$.

The elements of the domain $\{0,1\}^n$ are in 1-1 correspondence with subsets of $[n]$. Let $(P, \leqslant)$ be the poset with $P = 2^{[n]}$ and $\leqslant$ the subset - relation.

Exercise 4.129

Prove Lemma 4.212 below.

Hint: This lemma is 'the principle of inclusion - exclusion.'

Lemma 4.212 (Möbius inversion of the hypercube). *Let* $f : P \to \mathbb{R}$ *be a map and let* $g : P \to \mathbb{R}$ *be defined as follows*

$$g(x) = \sum_{y \leqslant x} f(y).$$

Then

$$f(x) = \sum_{y \leqslant x} g(y) \cdot (-1)^{n(x)-n(y)}, \qquad (4.23)$$

where $n(\cdot)$ *denotes the number of elements in the specified subset.*

The right hand - side of Equation 4.23 can be written as a multilinear polynomial in n variables $x_1, \cdots, x_n$: write $x \in P$ as $x = (x_1, \cdots, x_n)$ where each $x_i \in \{0, 1\}$. Then

$$(-1)^{n(x)} \quad = \quad (-1)^{\sum x_i} \quad = \quad \prod (-1)^{x_i} \quad = \quad \prod (1 - 2x_i).$$

In their paper Gotsman and Linial write $x \in P$ as

$$x = (x_1, \cdots, x_n) \quad \text{where} \quad x_i = \begin{cases} -1 & \text{if } i \in x \\ +1 & \text{if } i \notin x. \end{cases}$$

C. Gotsman and N. Linial, The equivalence of two problems on the cube, *Journal of Combinatorial Theory, Series A* **61** (1992), pp. 142–146.

and they rewrite (4.23) as

$$f(x) \quad = \quad \sum_{y \in P} \left(\alpha_y \cdot \prod_{i \in y} x_i \right) \quad = \quad \sum_{y \in P} \alpha_y \cdot (-1)^{n(x \cap y)}. \quad (4.24)$$

When $f : \{-1, +1\}^n \to \{-1, +1\}$ is a Boolean map then all of the 2^n coefficients satisfy $-1 \leqslant \alpha_x \leqslant +1$.

Gotsman and Linial call the coefficient α_x the Fourier transform of f at x.

Definition 4.213. The degree of a Boolean map $f : \{-1, +1\}^n \to \{-1, +1\}$ is

$$\delta(f) \quad = \quad \max\{ |x| \mid \alpha_x \neq 0 \}.$$

4.17.5   The equivalence theorem

In their paper Gotsman and Linial prove the equivalence theorem.

Let G be an induced subgraph of Q_n. Define

$$\Gamma(G) \quad = \quad \max \ \{ \Delta(G), \Delta(Q_n - G) \}.$$

Theorem 4.214 (The equivalence theorem). *Let $h : \mathbb{N} \to \mathbb{R}$ be a monotone map. The following two statements are both true or both false.*

1. if G is an induced subgraph of Q_n and $|V(G)| \neq 2^{n-1}$ then $\Gamma(G) \geqslant h(n)$

2. *any Boolean function* $f: 2^{[n]} \to \{-1, +1\}$ *satisfies* $s(f) > h(\delta(f))$.

Proof. Identify an induced subgraph G of Q_n with a Boolean function:

$$g(x) \quad = \quad \begin{cases} +1 & \text{if } x \in V(G) \\ -1 & \text{if } x \notin V(G). \end{cases}$$

Exercise 4.130

Show that

$$d_G(x) \quad = \quad n - s(g(x)).$$

Let $E(g) = \frac{1}{2^n} \cdot \sum_{x \in 2^{[n]}} g(x)$.

Exercise 4.131

Show that the statements in the lemma are equivalent with the following pair of statements.

1'. if g is a Boolean function and $E(g) \neq 0$ then there exists $x \in V(Q_n)$ with $s(g(x)) \leqslant n - h(n)$

2'. for any Boolean function f: if $s(f) < h(n)$ then $\delta(f) < n$.

The equivalence of 1.' and 2.' is shown as follows. Let

$$g(x) \quad = \quad f(x) \cdot \prod_{1}^{n} x_i.$$

Call the coefficients α_x ($x \in 2^{[n]}$) in (4.24), for the boolean functions f and g: $\hat{f}_x$ and $\hat{g}_x$.

f and g represent induced subgraphs with vertex sets that partition $V(Q_n)$.

Exercise 4.132

(a) Show that $s(g(x)) = n - s(f(x))$

(b) Show that the coefficients $\hat{f}_x$ and $\hat{g}_x$ of Equation (4.24) for the Boolean functions f and g satisfy

$$\hat{g}_x \quad = \quad \hat{f}_{\bar{x}} \qquad \text{where } \bar{x} = [n] - x$$

(c)
$$E(g) \quad = \quad \hat{g}_\varnothing \quad = \quad \hat{f}_{[n]}.$$

<u>1.' ⟹ 2.':</u>
Assume $\delta(f) = n$. Then $\hat{f}_{[n]} \neq 0$ — so — $E(g) \neq 0$. By 1.' there exists $x \in 2^{[n]}$ such that $s(g(x)) \leqslant n - h(n)$. This implies that there exists $x \in 2^{[n]}$ such that $s(f(x)) \geqslant h(n)$.

<u>2.' ⟹ 1.':</u>
Assume that for all $x \in 2^{[n]}$ $s(g(x)) > n - h(n)$. Then $s(f) < h(n)$. By 2.' $\delta(f) < n$ — and so —
$$\hat{f}_{[n]} \quad = \quad \hat{g}_\varnothing \quad = \quad E(g) \quad = \quad 0.$$

This proves the lemma.   □

We omit the proof of the following lemma. It was proved by Tal :

A. Tal, Properties and applications of Boolean function composition *Electronic Colloquium on Computational Complexity*, Report No. 163, 2012.

Lemma 4.215 (Tal). *The block sensitivity and degree of a Boolean functions satisfy*
$$bs(f) \quad \leqslant \quad \delta(f)^2.$$

We'll add a proof later on ... maybe ...

WE NOW PROVE THEOREM 4.208 (on Page 298).

Proof. In the equivalence theorem take $h(n) = \sqrt{n}$. Then the first item (1.) holds true since one of G or $Q_n - G$ has at least $2^{n-1} + 1$ vertices. We conclude
$$s(f) \quad \geqslant \quad \sqrt{\delta(f)}.$$

We obtain (by Tal's lemma) :
$$s(f)^4 \quad \geqslant \quad \delta(f)^2 \quad \geqslant \quad bs(f).$$

This proves Theorem 4.208.   □

4.17.6 Further reading

The sensitivity conjecture stems from this paper. The paper shows (Lemma 7) that $bs(f) \leqslant 2 \cdot \delta(f)^2$. (Tal's lemma removes the factor 2.)

N. Nisan and M. Szegedy, On the degree of Boolean functions as real polynomials, *Computational Complexity* **4** (1994), pp. 301–313.

The following paper makes a probabilistic approach.

J. Bourgain, J. Kahn, G. Kalai, Y. Katznelson and N. Linial, The influence of variables in product spaces, *Israel Journal of Mathematics* **77** (1992), pp. 55–64.

A LOT ABOUT EIGENVALUES and about interlacing techniques can be found in the following publications.

Andries E. Brouwer and Willem H. Haemers, *Spectra of Graphs*, Universitext, Springer, 2011.

Willem H. Haemers, Interlacing eigenvalues and graphs, *Linear Algebra and its Applications* 227/228, (1995), pp. 593–616.

The paper below is a classic on Möbius functions.

Gian - Carlo Rota, On the foundations of combinatorial theory: **I** Theory of Möbius functions, *Z. Wahrscheinlichkeitstheorie* **2** (1964), pp. 340–368.

4.18 Homomorphisms

Definition 4.216. Let G and H be two graphs. A homo-morphism

$$G \rightarrow H$$

is a map $h : V(G) \rightarrow V(H)$ with the property

$$\forall_{e \in E(G)} \quad h(e) \in E(H),$$

where — for a set $A \subseteq V(G)$ we write

$$h(A) = \{ h(a) \mid a \in A \}. \quad (4.25)$$

Thus, a homomorphism is a map that sends edges to edges.
133

[133] and dust to dust . . .

Exercise 4.133

Define $\preceq_{\text{hom}}$ as the quasi order defined on graphs by

$$G \preceq_{\text{hom}} H \quad \text{if there exists a homomorphism } G \to H.$$

Show that $\preceq_{\text{hom}}$ is not a well quasi–order.
Hint: Show that the sequence of odd cycles

$$C_3 , C_5 , C_7 , \cdots$$

is an infinite <u>antichain</u> —that is— no two elements are comparable under $\preceq_{\text{hom}}$. BTW, how about the even cycles?

Exercise 4.134

Show that for any graph G and $k \in \mathbb{N}$

$$G \to K_k \quad \Leftrightarrow \quad \chi(G) \leqslant k \tag{4.26}$$
$$K_k \to G \quad \Leftrightarrow \quad \omega(G) \geqslant k. \tag{4.27}$$

4.18.1 Retracts

Definition 4.217. Let G and H be graphs. The graph H is a <u>retract</u> of G if there exist homomorphisms [134]

$$\rho : G \to H \quad \text{and} \quad \gamma : H \to G \quad \text{such that} \quad \rho \circ \gamma = \text{id}_H$$

where id_H is the identity map $V(H) \to V(H)$.

When H is a retract of G then H is isomorphic to an induced subgraph of G. [135] Since there are homomorphisms in two directions,

[134] The maps ρ and γ are called the retraction and co-retraction.

[135] However a copy of H in G is not necessarily a retract of G. (When H is a retract then a proper coloring of H extends to a proper coloring of G.)

G and H have the same clique number, chromatic number and odd girth. A graph G retracts to K_k if and only if $\chi(G) = \omega(G) = k$.

For any graph H to check if there is a homomorphism $G \to H$ is polynomial when H is bipartite and it is NP-complete otherwise. It follows that, for any graph H, checking if H is a retract of a graph G is NP-complete, unless H is bipartite. The question whether a graph G has a homomorphism to itself which is not the identity is NP-complete.

4.18.2   Retracts in threshold graphs

Theorem 4.218. *Let G and H be threshold graphs. There exists a linear-time algorithm to check if H is a retract of G.*

Proof. Assume that H is a retract of G and let ρ and γ be the retraction and co-retraction.

Assume that G has a universal vertex, say x_1. Then H must have a universal vertex as well, since a retract of a connected graph is connected. Let y_1 be a universal vertex of H. Let $y_i = \rho(x_1)$. Since ρ is a homomorphism it preserves edges, and since x_1 is universal in G, ρ maps no other vertex of G to y_i. Notice also that $\gamma(y_i) = x_1$ since $\rho \circ \gamma = id_H$ and ρ maps no other vertex to y_i.

Assume that $y_i \neq y_1$. Let $\gamma(y_1) = x_\ell$. Then $x_\ell \neq x_1$ since γ preserves edges and so

$$\{y_1, y_i\} \in E(H) \quad \Rightarrow$$
$$\{\gamma(y_1), \gamma(y_i)\} = \{x_\ell, x_1\} \in E(G) \quad \Rightarrow \quad x_\ell \neq x_1.$$

Furthermore, since y_1 is universal, γ maps no other vertex of H to x_ℓ. Of course, since $\rho \circ \gamma = id_H$, $\rho(x_\ell) = y_1$.

We claim that y_i is universal in H, and therefore exchangeable with y_1. Assume not and let $y_s \in V(H)$ be another vertex of H not adjacent to y_i. Let $\gamma(y_s) = x_p$. Then $x_p \neq x_1$ since $\rho \circ \gamma = id_H$ and $\rho(x_1) = y_i \neq y_s$. Now, since ρ is a homomorphism,

$$\{x_1, x_p\} \in E(G) \quad \Rightarrow \quad \{\rho(x_1), \rho(x_p)\} = \{y_i, y_s\} \in E(H),$$

which is a contradiction. Therefore, we may assume that $y_i = y_1$.
— That is — from now on we assume that

$$\rho(x_1) = y_1 \quad \text{and} \quad \gamma(y_1) = x_1.$$

This proves that, when G is connected then H is a retract of G if and only if H — y_1 is a retract of G — x_1. By the way, notice that if $|V(H)| = 1$ then H can be a retract of G only if G is an independent set, so this case is easy to check.

Finally, assume that G is not connected. Since G has no induced $2K_2$, all components, except possibly one, have only one vertex. The number of components of H can be at most equal to the number of components of G, since ρ maps components in G to components of H, and $\rho \circ \gamma = id_H$, and so any two components of H are mapped by γ to different components of G.

First assume that H is also disconnected. Let $x_1, \ldots, x_a$ be the isolated vertices of G and let $y_1, \ldots, y_b$ be the isolated vertices of H. Let $\rho(x_i) = y_i$ and $\gamma(y_i) = x_i$ for $i \in \{1, \ldots, b\}$ and let $\rho(x_{b+1}) = \cdots = \rho(x_a) = y_b$. Now, H is a retract of G if and only if $H - \{x_1, \ldots, x_b\}$ is a retract of $G - \{x_1, \ldots, x_a\}$.

If H is connected, with at least two vertices, then let y_1 be a universal vertex and let $\rho(x_1) = \cdots = \rho(x_a) = y_1$. If H is a retract of G then G must have exactly one component with at least two vertices, since G is a threshold graph and ρ is a homomorphism. Let x_u be the universal vertex of that component and define $\rho(x_u) = y_1$ and $\gamma(y_1) = x_u$. In this case, H is a retract if and only if $H - y_1$ is a retract of $G - \{x_1, \ldots, x_a, x_u\}$.

An elimination ordering, which eliminates successive isolated and universal vertices in a threshold graph, can be obtained in linear time.

This proves the theorem. □

4.18.3 Retracts in cographs

In this section we show that the retract - problem is NP-complete on cographs.

RECALL THAT A GRAPH G IS PERFECT WHEN $\omega(G') = \chi(G')$
FOR EVERY INDUCED SUBGRAPH G' OF G. By the perfect graph
theorem a graph is perfect if and only if it has no odd hole or odd
antihole. This implies that cographs are perfect. Perfect graphs are
recognizable in polynomial time. For a graph G, when $\omega(G) = \chi(G)$ one can compute this value in polynomial time via Lovász
theta function.

Lemma 4.219. *Assume that* $\omega(H) = \chi(H)$. *There is a homomorphism* $G \to H$ *if and only if* $\chi(G) \leqslant \omega(H)$.

Proof. Write $\omega = \omega(H) = \chi(H)$. First assume that there is a
homomorphism $\phi : G \to H$. There is a homomorphism $f : H \to K_\omega$
since H is ω-colorable. Then $f \circ \phi : G \to K_\omega$ is a homomorphism,
and so G has an ω-coloring. This implies that $\chi(G) \leqslant \omega$.

Assume $\chi(G) \leqslant \omega$. There is a homomorphism $G \to K_k$, where
$k = \chi(G)$. Since K_k is an induced subgraph of H, there is also a
homomorphism $K_k \to H$. This implies that G is homomorphic to H
— ie — $G \to H$.

This proves the lemma. $\square$

Corollary 4.220. *When* G *and* H *are perfect one can check
in polynomial time whether there is a homomorphism* $G \to H$.

RETRACTS — LIKE GENERAL HOMOMORPHISMS — CONSTITUTE A TRANSITIVE RELATION. We provide a short proof of this
for completeness sake.

Lemma 4.221. *Let* A *be a retract of* G *and let* B *be a retract of* A.
Then B *is a retract of* G.

Proof. Let ρ_1 and γ_1 be a retraction and co-retraction from G to
A and let ρ_2 and γ_2 be a retraction and co-retraction from A to

B. Since all four maps ρ_1, ρ_2, γ_1 and γ_2 are homomorphisms, the following two maps are homomorphisms as well.

$$\rho_2 \circ \rho_1 : G \to B \quad \text{and} \quad \gamma_1 \circ \gamma_2 : B \to G. \tag{4.28}$$

Furthermore,

$$(\rho_2 \circ \rho_1) \circ (\gamma_1 \circ \gamma_2) = \rho_2 \circ id_A \circ \gamma_2 = \rho_2 \circ \gamma_2 = id_B. \tag{4.29}$$

This proves that B is a retract of G. $\qquad\qquad\square$

Throughout the remainder of this section it is assumed that G and H are cographs. Note that, using the cotree, $\omega(G)$ and $\chi(G)$ can be computed in linear time when G is a cograph.

Lemma 4.222. *Assume* H *is disconnected; denote the components of* H *as*

$$H_1 \quad \cdots \quad H_t.$$

Assume that H *is a retract of a graph* G. *Then there is an ordering of the components of* G, *say* $G_1, \cdots, G_s$ *such that*

(a) $s \geqslant t$, *and*

(b) G_i *retracts to* H_i, *for every* $i \in \{1, \ldots, t\}$, *and*

(c) *for every* $j \in \{t+1, \ldots, s\}$, *there is a homomorphism* $G_j \to H$.

Proof. No connected graph has a disconnected retract since the homomorphic image of a connected graph is connected. To see that, notice that a homomorphism $\phi : G \to H$ is a vertex coloring of G, where the vertices of H represent colors. By that we mean that, for each $v \in V(H)$, the pre-image $\phi^{-1}(v)$ is an independent set in G or $\emptyset$. One obtains the image $\phi(G)$ by identifying vertices in G that receive the same color. When G is connected, this 'quotient graph' on the color classes is also connected, which is easy to prove by means of contradiction.

Assume that G retracts to H. Then we may assume that $H_1, \ldots, H_t$ are induced subgraphs of components $G_1, \ldots, G_t$ of G and that each G_i retracts to H_i. For the remaining components G_j, where $j > t$, there is then a homomorphisms $G_j \to H$.

Notice that, for $j > t$, we can check if there is a homomorphism $G_j \to H$ by checking if $G_j \oplus H_k$ retracts to H_k for some $1 \leqslant k \leqslant t$ — or, equivalently (since cographs are perfect) — if $\omega(G_j) \leqslant \omega(H_k)$ for some $1 \leqslant k \leqslant t$. $\qquad\qquad\qquad\qquad\qquad\qquad\qquad\qquad$ □

Remark 4.223. Assume that we are given, for each pair G_i and H_j whether G_i retracts to H_j or not. Then, to check if G retracts to H, we may consider a bipartite graph B defined as follows. One color class of B has the components of G as vertices and the other color class has the components of H as vertices. There is an edge between G_i and H_j whenever G_i retracts to H_j. To check if G retracts to H, we can let an algorithm compute a maximum matching in B. There is a retraction only if the matching exhausts all components of H and if $\omega(G) = \omega(H)$.

A cocomponent of a graph G is a subset of vertices which induces a component of the complement $\bar{G}$.

Lemma 4.224. *Assume* G *is connected and assume that* G *retracts to* H. *Then* H *is also connected. Let* $G_1, \cdots, G_t$ *be the subgraphs of* G *induced by the cocomponents of* G. *There is a partition of the cocomponents of* H *such that the subgraphs of* H *induced by the parts of the partition can be ordered* $H_1, \cdots, H_t$ *such that* G_i *retracts to* H_i *for* $i \in \{1, \cdots, t\}$.

Proof. Every subgraph G_i of G, induced by a cocomponent, some subgraphs induced by cocomponents of H. Thus the parts of $V(H)$ that are the images of the subgraphs induced by cocomponents of G form a partition of the cocomponents of H. $\qquad\qquad$ □

Theorem 4.225. *Let* G *and* H *be cographs. The problem to decide whether* H *is a retract of* G *is* NP-*complete.*

Proof. We reduce the 3-partition problem to the retract problem on cotrees. The 3-partition problem is the following. Let m and B be integers. Let S be a multiset of $3m$ positive integers, $a_1, \ldots, a_{3m}$. Determine if there is a partition of S into m subsets $S_1, \ldots, S_m$, such that the sum of the numbers in each subset is B. Without loss of generality we assume that each number is strictly between $B/4$ and $B/2$, which guarantees that in a solution each subset contains exactly three numbers that add up to B.

The 3-partition problem is strongly NP-complete, that is, the problem remains NP-complete when all the numbers in the input are represented in unary.

In our reduction, the cotree for the graph H has a root which is labeled as a join-node $\otimes$. The root has $3m$ children, one for each number a_i. For simplicity we refer to the children as a_i, $i \in \{1, \ldots, 3m\}$. Each child a_i has a union node $\oplus$ as the root. The root of each a_i-child has two children, one is a single leaf and the other is a join-node $\otimes$ with a_i leaves. This ends the description of H.

The cotree for the graph G has a join-node $\otimes$ as a root and this has m children. The idea is that each child corresponds with one set of a 3-partition of S. The subtrees for all the children are identical. It has a union-node $\oplus$ as the root. Consider all triples $\{i, j, k\}$ for which $a_i + a_j + a_k = B$. For each such triple create one child, which is the join of three cotrees, one for a_i, one for a_j and one for a_k in the triple. The subtree for a_i is a union of two subtrees. As in the cotree for the pattern H, one subtree is a single leaf, and the other subtree is the join of a_i leaves. The other two subtrees, for the numbers a_j and a_k in the triple are similar.

Let T_H and T_G be the cotrees for H and G as constructed above. Say T_H and T_G have roots r_H and r_G. When the graph H is a retract of G then the a_i-children of r_H are partitioned into triples, such that there is a bijection between these triples, say $\{a_i, a_j, a_k\}$ and a branch in the cotree of G. Each $\oplus$-node which is the root of a child of r_G must have exactly one $\{a_i, a_j, a_k\}$-child that corresponds with the triple. Notice that, by the construction, all subgraphs induced by remaining components of the $\oplus$-node have maximal cliques of size B. Therefore, all other children of the $\oplus$-node are homomorphic to the one child which corresponds to the triple $\{a_i, a_j, a_k\}$.

It now follows from the Lemma above that there is a 3-partition if and only if the graph H is a retract of G.

This completes the proof. $\square$

4.19   Products

Let G and H be graphs. The categorical product (or tensor product) is a graph denoted as $G \times H$. [136] The vertices of $G \times H$ are

$$V(G \times H) \quad = \quad \{\, (g, h) \mid g \in V(G) \quad \text{and} \quad h \in V(H) \,\}.$$

Two vertices — say (g_1, h_1) and (g_2, h_2) — are adjacent if

$$\{g_1, g_2\} \in E(G) \quad \text{and} \quad \{h_1, h_2\} \in E(H).$$

[136] The categorical product also goes by the name of 'tensor product' or 'direct product', 'Kronecker product' and more. This is just one way to define a graph product.

Hedetniemi made the following CONJECTURE (some 50 years ago). For any two graphs G and H

$$\chi(G \times H) \quad = \quad \min\{\chi(G), \chi(H).\} \tag{4.30}$$

The right hand - side is an upperbound. To see that let f be a coloring of G. Let f' be defined as

$$\text{for } g \in V(G) \text{ and } h \in V(H): \quad f'(g, h) \quad = \quad f(g).$$

Then f' is a proper coloring of $G \times H$.

The 'fractional version' of Hedetniemi's conjecture is true: X. Zu, *The fractional version of Hedetniemi's conjecture is true*, European Journal of Combinatorics **32** (2011), pp. 1168–1175.

Yaroslav Shitov produced a COUNTEREXAMPLE to the conjecture in 2019.

Y. Shitov. Counterexamples to Hedetniemi's conjecture. Manuscript on ArXiv: 1905:02167, 2019.

Theorem 4.226. *When G and H are perfect then* (4.30) *holds true.*

Proof. CLEARLY [137]

$$\omega(G \times H) \quad \geqslant \quad \min\{\omega(G), \omega(H)\}.$$

[137] Exercise !

When we assume that G and H are perfect then

$$\chi(G) = \omega(G) \quad \text{and} \quad \chi(H) = \omega(H).$$

The claim easily follows from the following observation.

$$\chi(G \times H) \geq \omega(G \times H) \geq \min\{\omega(G), \omega(H)\} = $$
$$\min\{\chi(G), \chi(H)\} \geq \chi(G \times H).$$

This proves the theorem. □

4.19.1 Categorical products of cographs

MUCH LESS IS KNOWN ABOUT THE INDEPENDENCE NUMBER IN THE CATEGORICAL PRODUCT OF GRAPHS. Clearly we have that [138]

$$\alpha(G \times H) \geq \max\{\alpha(G) \cdot |V(H), \alpha(H) \cdot |V(G)|, \} \quad (4.31)$$

but this lower bound is not sharp not even for threshold graphs. [139]

Cographs are perfect. But the product of two cographs is not necessarily perfect. As an example let G be isomorphic to the paw (see Figure 2.8 on Page 80) and let H be isomorphic to K_3. Then G × H contains C_5 as an induced subgraph. [140]

Ravindra and Parthasarathy showed that G × H is perfect if and only if one of the following holds. [141]

1. G or H is bipartite

2. G and H contain no odd holes and no paws.

[138] Exercise !

[139] P. Jha and S. Klavžar, *Independence in direct - product graphs*, Ars Combinatoria **50**, 1998.

[140] Check !

[141] G. Ravindra and K. Parthasarathy, *Perfect product graphs*, Discrete Mathematics **20** (1977), pp. 177–186.

Exercise 4.135

(a) Let G and H be complete multipartite. Then G × H is perfect.

(b) If G and H are complete multipartite then

$$\alpha(G \times H) \quad = \quad \max\{\,\alpha(G) \cdot |V(H)|,\ \alpha(H) \cdot |V(G)|\,\}.$$

Exercise 4.136

Let G and H be cographs and assume that G is diconnected. Say $G = G_1 \oplus G_2$ (G is the union of G_1 and G_2). Show that

$$\alpha(G \times H) \quad = \quad \alpha(G_1 \times H) + \alpha(G_2 \times H).$$

Exercise 4.137

Let G and H be cographs and assume that both are connected. Say $G = G_1 \otimes G_2$ and $H = H_1 \otimes H_2$ (G and H are joins of the constituents). Show that

$$\alpha(G \times H) \quad = \quad \max\{\,\alpha(G_1 \times H),\ \alpha(G_2 \times H),\ \alpha(G \times H_1),\ \alpha(G \times H_2)\,\}.$$

WE CAN NOW LEAVE THE PROOF OF THE FOLLOWING THEO-REM AS AN EXERCISE.

Theorem 4.227. *There exists an* $O(n^2)$ *algorithm to compute* $\alpha(G \times H)$ *when* G *and* H *are cographs.*

Exercise 4.138

Show that there is a polynomial - time algorithm to compute the independence number of $G \times H$ when G and H are splitgraphs.

4.19.2 Tensor capacity

Definition 4.228. The independence ratio of a graph G is defined as

$$r(G) \quad = \quad \frac{\alpha(G)}{|V(G)|}.$$

Exercise 4.139

Show that
$$r(G \times H) \quad \geqslant \quad \max\{r(G), r(H)\}.$$

HINT: Use (4.31).

WRITE G^k FOR $G \times \cdots \times G$ where G is $k-1$ times multiplied by itself. Notice that $r(G^k)$ is non-decreasing and it is at most 1. Therefore $\lim_{k \to \infty} r(G^k)$ exists. Call this limit the tensor capacity of the graph.

Definition 4.229. Let G be a graph. The <u>tensor capacity</u> of G is
$$\Theta(G) \quad = \quad \lim_{k \to \infty} r(G^k).$$

It can be shown that the computation of the tensor capacity is NP-complete.

Let G be a graph. Define
$$a(G) \quad = \quad \max \frac{|I|}{|I| + |N(I)|}$$

where I varies over the independent sets in G. Define
$$a^*(G) = \begin{cases} a(G) & \text{if } a(G) \leqslant 1/2 \\ 1 & \text{if } a(G) > 1/2. \end{cases}$$

Tóth proved the following theorem. [142]

Theorem 4.230.
$$\Theta(G) \quad = \quad a^*(G).$$

Equivalently for any graph
$$a^*(G^2) \quad = \quad a^*(G).$$

Theorem 4.231. *There exists a polynomial - time algorithm to compute the tensor capacity for cographs.*

[142] Á. Tóth, *Answer to a question of Alon and Lubetzky about the ultimate categorical independence ratio.* Manuscript on arXiv: 1112.6172, 2011.

Let G and H be graphs. Tóth showed that
$$\Theta(G \oplus H) = \Theta(G \times H) = \max\{\Theta(G), \Theta(H)\}.$$

Proof. By Tóth's result, it is sufficient to show that $a(G)$ can be computed.

Consider a cotree. For each node in the cotree the algorithm computes a table with numbers $\ell(k)$

$$\ell(k) \quad = \quad \min\{|N(I)| \mid I \text{ is an independent set and } |I| = k\}.$$

The value $a(G)$ is then obtained from the table at the root as

$$a(G) \quad = \quad \max_k \frac{k}{k + \ell(k)}.$$

Assume that G is disconnected — say $G = G_1 \oplus G_2$. Let ℓ_1 and ℓ_2 denote the tables for G_1 and G_2. Then

$$\ell(k) \quad = \quad \min\{\ell_1(k_1) + \ell_2(k_2) \mid k_1 + k_2 = k\}.$$

Let G be connected — say $G = G_1 \otimes G_2$. In that cae we have

$$\ell(k) \quad = \quad \min\{\ell_1(k) + |V(G_2)|, |V(G_1)| + \ell_2(k)\}.$$

This proves the theorem. $\square$

Exercise 4.140

Show that there is an $O^*(3^{n/3})$ algorithm to compute the tensor capacity of a graph.

HINT: Use Moon and Moser.

Remark 4.232. It can be shown that the tensor capacity can be computed in time $O(3^{k+1} \cdot n^3)$ for graphs of treewidth at most k.

It is NP-complete to determine $\alpha(G \times K_4)$ when G is a planar graph of maximal degree 3.

4.19.3   Cartesian products

THE CARTESIAN PRODUCT $G \square H$ of two graphs G and H is the graph with vertex set $V(G) \times V(H)$ and (g_1, h_1) adjacent to (g_2, h_2) if

$$g_1 = g_2 \quad \text{and} \quad \{h_1, h_2\} \subset E(H) \quad \text{or}$$
$$h_1 = h_2 \quad \text{and} \quad \{g_1, g_2\} \in E(G).$$

Independence domination

Definition 4.233. Let G be a graph. A set $B \subseteq V$ <u>dominates</u> a set $A \subseteq V$ when

$$A \quad \subseteq \quad \bigcup_{x \in B} N[x].$$

The minimal cardinality of a set B that dominates a set A is denoted as $\gamma(A)$.

Definition 4.234. The <u>independence domination number</u> of a graph G is

$$\gamma^i(G) \quad = \quad \max \{ \gamma(A) \mid A \text{ an independent set} \}.$$

CLEARLY — for any graph $\gamma \geqslant \gamma^i$.

Vizing's CONJECTURE states

$$\gamma(G \square H) \quad \geqslant \quad \gamma(G) \cdot \gamma(H).$$

Aharoni and Szabó proved in 2009 that Vizing's conjecture holds true for chordal graphs.   — Furthermore — they show that for all graphs G and H

$$\gamma(G \square H) \quad \geqslant \quad \gamma^i(G) \cdot \gamma(H) \quad \text{and} \quad \gamma^i(G \square H) \quad \geqslant \quad \gamma^i(G) \cdot \gamma^i(H).$$

Here $\gamma = \gamma(G)$ is the <u>domination number</u> of G: it is the smallest cardinality of a dominating set D — a set D that satisfies $N[x] \cap D \neq \varnothing$ for every $x \in V$.

A graph is <u>chordal</u> if it has no induced cycle of length more than 3. Computing γ for chordal graphs is NP-complete.

PROGRESS towards proving the conjecture was made by Suen and Tarr in 2012. They proved

$$\gamma(G \square H) \;\geqslant\; \frac{1}{2} \cdot \gamma(G) \cdot \gamma(H) \;+\; \min\{\gamma(G), \gamma(H)\}.$$

4.19.4   Independence domination in cographs

WHEN G IS A COGRAPH it is either the join or the union of two cographs — say

$$G = G_1 \otimes G_2 \quad \text{or} \quad G = G_1 \oplus G_2.$$

Exercise 4.141

When G is a cograph with at least two vertices then

$$\gamma(G) \;=\; \begin{cases} \min\{\gamma(G_1), \gamma(G_2), 2\} & \text{if } G = G_1 \otimes G_2 \\ \gamma(G_1) + \gamma(G_2) & \text{if } G = G_1 \oplus G_2. \end{cases}$$

Exercise 4.142

When G is a cograph then $\gamma^i(G)$ is the number of components of G.

Exercise 4.143

Let $k \in \mathbb{N}$. Design a polynomial - time algorithm to compute γ^i for graphs of rankwidth $\leqslant k$.

Exercise 4.144

Show that there is a polynomial - time algorithm to compute γ^i for permutation graphs.

Wing-Kai Hon, T. Kloks, H. Liu, S. Poon and Yue-Li Wang, *On independence domination*. Manuscript on arXiv: 1304.6450, 2013.

4.19.5 $\theta_e(K_n \times K_n)$

EVERY GRAPH IS THE INTERSECTION GRAPH OF A COLLEC-
TION OF SUBSETS OF A SET U. By that we mean that every
vertex is represented by a subset of U and two vertices are adjacent
precisely when the subsets have a nonempty intersection.

For example interval graphs
are intersection graphs.

Definition 4.235. For a graph G let θ_e denote the minimal size
of a set U such that G is the intersection graph of a collection of
subsets of U.

For the reason given below the parameter θ_e is called the <u>edge
clique cover - number</u> of the graph.

Show that θ_e is finite for
any graph. **Hint:** Number
the edges of the graph
$e_1, \cdots, e_m$. For each vertex x let $S_x = \{ j \mid x \in e_j \}$.

Exercise 4.145

Show that θ_e is the minimal size of a set of cliques that has the
property that every edge is contained in at least one of them.

The tensor product $K_n \times K_n$

The tensor product $K_n \times K_n$ has the following set of vertices

$$V \quad = \quad \{ (i, j) \mid i \in [n] \quad \text{and} \quad j \in [n] \}.$$

Two vertices (i, j) and (k, ℓ) are adjacent if and only if

$$i \neq k \quad \text{and} \quad j \neq \ell.$$

Exercise 4.146

Show that any graph satisfies

$$\theta_e \quad \geqslant \quad m / \binom{\omega}{2}.$$

Corollary 4.236.

$$\theta_e(K_n \times K_n) \;\geqslant\; n(n-1).$$

Exercise 4.147

Let p be prime and let $u \leqslant p$. Show that

$$\theta_e(K_p \times K_u) \;=\; p(p-1).$$

The following theorem characterizes those $n \in \mathbb{N}$ for which $K_n \times K_n$ has an edge clique - cover with $n(n-1)$ cliques.

Theorem 4.237. $\theta_e(K_n \times K_n) = n(n-1)$ *if and only if there exists a projective plane of order* n.

We omit the proof. The theorem is reminiscent of a result of De Bruin and Erdős concerning $\theta_e(K_n)$.

Wing-Kai Hon, Ton Kloks, Hsiang-Hsuan Liu and Yue-Li Wang, *Edge clique - covers of the tensor product*, Theoretical Computer Science **607** (2015), pp. 68–74.

Definition 4.238. Let $n \in \mathbb{N}$. A <u>projective plane</u> of order n is a set of $n^2 + n + 1$ <u>points</u> and $n^2 + n + 1$ <u>lines</u> such that [143]

P1. every line has $n+1$ points on it

P2. every point is on $n+1$ lines

P3. any two lines intersect in exactly one point

P4. any two points lie on exactly one line

[143] Lines are sets of points. We say that a point lies on a line if the line contains it.

The Fano plane is a projective plane of order two. A projective plane of order 6 does not exist. The case $n = 10$ was ruled out by computer calculations. The existence of a projective plane of order 12 is open. There exists a projective plane of order n when n is a prime power.

Corollary 4.239. *For every* n *which is the power of a prime number*

$$\theta_e(K_n \times K_n) \quad = \quad n(n-1).$$

Exercise 4.148

For $n \geqslant 2$

$$\theta_e(K_n \times K_n) \quad \leqslant \quad (2n-1) \cdot (2n-2).$$

HINT: Let p be the smallest prime $\geqslant n$. Bertrand's postulate says that $p \leqslant 2n - 1$.

Remark 4.240. When the Riemann hypothesis holds true then

$$\lim_{n \to \infty} \frac{\theta_e(K_n \times K_n)}{n(n-1)} \quad = \quad 1.$$

4.20 Outerplanar Graphs

An embedding of a graph G in the plane is a drawing of G such that no two edges intersect. A plane graph is already embedded.

The maximal regions — bounded by the edges of the graph — are called <u>faces</u>. The unbounded region is unique, and it is called the <u>outerface</u> .

Definition 4.241. A planar graph is <u>outerplanar</u> if it can be embedded in the plane so that all its vertices lie on the same face. Customarily, this face is called the exterior.

Figure 4.29: A MOP

Exercise 4.149

Show that the class of outerplanar graphs is closed under taking minors. The obstruction set is

$$\mathcal{O} = \{K_4, K_{2,3}\}.$$

Exercise 4.150

A recursive definition of a MOP is the following. [144]

(i) A graph consisting of a single edge is a MOP,

(ii) If G is a MOP, then a new MOP is constructed by adding a vertex and making it adjacent to the endpoints of an edge that is not a minimal separator of G.

A <u>maximal outerplanar graph</u> is an outerplanar graph with an inclusion–wise maximal set of edges. — Thus — adding an edge destroys the outerplanarity.

Lemma 4.242. *Any outerplanar graph has treewidth at most 2.*

Proof. Any outerplanar graph embeds in a MOP. A maximal outerplanar graph has treewidth at most two.   □

4.20.1   k – Outerplanar Graphs

A parametrization of the class of all planar graphs is obtained by partitioning its vertices into outerplanar layers.

Definition 4.243. Let G be a plane graph. Its <u>layers</u>, say

$$L_1, L_2, \cdots$$

form a partition of $V(G)$ where L_i is the set of vertices in the outerface of

$$G - \bigcup_{j=1}^{i-1} L_j.$$

The graph G is <u>k–outerplanar</u> if it has a plane embedding with at most k nonempty layers. [145]

Bodlaender proved the following generalization of Lemma 4.242.

Lemma 4.244. *A k–outerplanar graph has treewidth* $3k-1$. [146]

[144] Show that every MOP is Hamiltonian.

[145] Computing the (smallest) outerplanarity of a graph is NP-complete.

[146] See Figure 4.29. Contract the outerface to a MOP, such that each component of the remaining layers is contained in a triangle. Continue this process for the remaining layers. This adds one triangle per layer and so, the clique number is bounded by 3k.

4.20.2 Courcelle's Theorem

Courcelle proved in 1990 the following theorem.

Theorem 4.245. *Any problem that can be formulated in* MS_2 *can be solved in linear time for graphs of bounded treewidth.*

Courcelle's theorem — as above — is based on Bodlaender's linear–time algorithm to recognize graphs of treewidth at most k. The class of graphs that have treewidth at most k is minor–closed and characterized by a finite obstruction set — say $\mathcal{T}_k$. This implies that bounded treewidth can be formulated in monadic second–order logic. By Theorem 4.245, $\mathrm{tw}(G) \leqslant k$ can be tested in linear time for all $k \in \mathbb{N}$. [147]

[147] Bodlaender's algorithm constructs an embedding; for that, the obstruction set is not needed.

4.20.3 Approximations for Planar Graphs

Baker showed that many optimization problems on planar graphs can be approximated — to an arbitrary degree of accuracy — by a linear–time algorithm.

THIS ELEGANT METHOD is best explained via an example.

4.20.4 Independent Set in Planar Graphs

To compute α for planar graphs is NP-complete. [148] Baker's method provides an efficient approximation scheme.

[148] On the other hand, to compute $\omega(G)$ is polynomial.

Theorem 4.246. *For every* $k \in \mathbb{N}$ *there exists a linear–time algorithm that approximates* $\alpha(G)$ *— for planar graphs — within a factor* $\frac{k}{k+1}$.

Proof. Let G be a plane graph. Number its layers consecutively as

$$L_1, L_2, \cdots \quad \text{where } L_1 \text{ is the outerface.}$$

For $1 \leqslant i \leqslant k$ define the graph G_i as the graph induced by

$$\bigcup_{j \neq i \bmod k} L_j \, .$$

The k graphs G_i are k-outerplanar since the missing layers form separators. Thus — by Theorem 4.245 — the independence numbers $\alpha(G_i)$ can be computed in linear time. [149]

Let M be a maximum independent set in G. Then

$$\alpha(G) = \sum_i L_i \cap M \tag{4.32}$$

$$= \frac{1}{k-1} \cdot \sum_{i=1}^{k} \sum_{j \neq i \bmod k} L_j \cap M \tag{4.33}$$

$$\leqslant \frac{1}{k-1} \cdot \sum_{i=1}^{k} \alpha(G_i) \tag{4.34}$$

$$\leqslant \frac{1}{k-1} \cdot k \cdot \max\{\, \alpha(G_i) \mid 1 \leqslant i \leqslant k \,\} \tag{4.35}$$

$$\Rightarrow \max_i \alpha(G_i) \geqslant \frac{k-1}{k} \cdot \alpha(G). \tag{4.36}$$

This proves the theorem. [150]   □

[149] The problem of the independence number can be formulated in monadic second-order logic.

[150] Dash it all! I should've started with $k + 1$! I *always* do that!

4.21 Graph isomorphism

COMING SOON!

We regret that wo can not present the beautiful graph isomorphism test of Láeló Babai in this book.

Must - reads on graph isomorphism

Grohe, M. and D. Neuen, Recent advances on the graph isomorphism problem. Manuscript on ArXiv: 2011.01366, 2021.

Bibliography

[1] Alipour, S. and A. Jafari, *Upperbounds for domination number of graphs using Turán's and Lovász local lemma*. Manuscript on ArXiv: 1803.04031, 2018.

[2] Alon, N., R. Duke, H. Lefmann, V. Rödl and R. Yuster, *Algorithmic aspects of the regularity lemma*. Journal of Algorithms **16** (1994), pp. 80–109.

[3] Alon, N., R. Yuster and U. Zwick, *Finding and counting given length cycles*. Algorithmica **17** (1997), pp. 209–223.

[4] Arnborg, S., D. Corneil, and A. Proskurowski, *Complexity of finding embeddings in a k-tree*. SIAM Journal on Algebraic and Discrete Methods **8** (1987), pp. 277–284.

[5] Backus, J., *Can programming be liberated from the Von Neumann style? A functional style and its algebra of programs*. Communications of the ACM **21** (1978), pp. 613–641.

[6] Baker, B., *Approximation algorithms for NP-complete problems on planar graphs*. Journal of the ACM **41** (1994), pp. 153–180.

[7] Bell, R., *Dynamic Programming*. Dover Books on Computer Science, 2003.

[8] Berge, C., *Two theorems in graphs*. Proc. Nat. Acad. Sci. **43** (1957), pp. 842–844.

[9] Bouchet, A., *Reducing prime graphs and recognizing circle graphs*. Combinatorica **7** (1987), pp. 243–254.

© The Author(s), under exclusive license to Springer Nature Singapore Pte Ltd. 2022
T. Kloks, M. Xiao, *A Guide to Graph Algorithms*, https://doi.org/10.1007/978-981-16-6350-5

[10] Blokhuis, A., T. Kloks and H. Wilbrink, *A class of graphs containing the polar spaces*. European Journal of Combinatorics **7** (1986), pp. 105–114.

[11] Booth, K. and G. Lueker, *Testing for the consecutive ones property, interval graphs, and graph planarity using* PQ-*tree algorithms*. Journal of Computer and System Sciences **13** (1976), pp. 335–379.

[12] Bose, R., *Strongly regular graphs, partial geometries and partially balanced designs*. Pacific Journal of Mathematics **13** (1963) pp. 389–419.

[13] Broersma, H., T. Kloks, D. Kratsch and H. Müller, *Independent sets in asteroidal triple-free graphs*. SIAM Journal on Discrete Mathematics **12** (1999), pp. 276–287.

[14] Bron, C. and J. Kerbosch, *Algorithm 457: finding all cliques of an undirected graph*. Communications of the Association for Computing Machinery, ACM **16** (1973) pp. 575–577.

[15] Chudnovsky, M. and P. Seymour, *Claw-free graphs. III. Circular interval graphs*. Journal of Combinatorial Theory, Series B **98** (2008), pp. 812–834.

[16] Cook, S., *The complexity of theorem-proving procedures*. Third Annual ACM Symposium on Theory of Computing, Association for Computing Machinery, New York, 1971, pp. 151–158.

[17] Corneil, D., Y. Perl and L. Stewart, *A linear time recognition algorithm for cographs*. SIAM Journal on Computing **14** (1985), pp. 926–934.

[18] Courcelle, B., and J. Engelfriet, *Graph structure and monadic second-order logic – A language theoretic approch*. Cambridge University Press, 2012.

[19] Dirac, G., *On rigid circuit graphs*. Congress Report University Hamburg **25** (1961), pp. 71–76.

[20] Dijkstra, E., *Guarded commands, non-determinacy and formal derivation of programs*. Note EWD 472.

[21] Dijkstra, E., *A discipline of programming*. Prentice-Hall, 1976.

[22] Edmonds, J., *Path, trees and flowers*. The Canadian Journal of Mathematics **17** (1965), pp. 449–467.

[23] Eppstein, D., M. Löffler and D. Strash, *Listing all maximal cliques in sparse graphs in near-optimal time*. Manuscript on ArXiV: 1006.5440v1, 2010

[24] Faenza, Y., G. Oriolo and G. Stauffer, *An algorithmic decomposition of claw-free graphs leading to an $O(n^3)$-algorithm for the weighted stable set problem*. Proceedings of the 22nd Annual ACM-SIAM Symposium on Discrete Algorithms, 2011.

[25] Fox, J. and L. M. Lovász, *A tight lower bound for Szemerédi's regularity lemma*. Manuscript on ArXiv: 1403.1768, 2014.

[26] Frederickson, G., *Fast algorithms for shortest paths in planar graphs, with applications*. Technical Report 84-486, Purdue University, 1986.

[27] le Gall, F., *Powers of tensors and fast matrix multiplication*. Manuscript on ArXiv: 1401.7714, 2014.

[28] Garey, M. and D. Johnson, *Computers and Intractibility: A guide to the theory of NP-completeness*. W. Freeman.

[29] Gioan, E., C. Paul, M. Tedder and D. Corneil, *Practical and efficient circle graph recognition*. Algorithmica **69** (2014), pp. 759–788.

[30] Harary, F., *Graph Theory*. Addison-Wesley, 1972.

[31] Higman, G., *Ordering by divisibility in abstract algebras*. Proceedings of the London Mathematical Society **2** (1952), pp. 326–336.

[32] Hung, L. and T. Kloks, *k-cographs are Kruskalian*. Chicago Journal of Theoretical Computer Science, 2011.

[33] Hung, L., T. Kloks and F. Villaamil, *Black-and-white threshold graphs*. ArXiV:1104-3917.

[34] Kahn, A., *Topological sorting of large networks*. Communications of the ACM, **5** (1962), pp. 558–562.

[35] Kloks, T., *Treewidth of circle graphs*. Utrecht University, RUU-CS-1993-12, 1993.

[36] Kloks, T., *Treewidth*. Springer-Verlag, Lecture Notes in Computer Science 842, 1994.

[37] Kloks, T., D. Kratsch and J. Spinrad, *Treewidth and pathwidth of cocomparability graphs of bounded dimension*. Technische Universiteit Eindhoven, Computing Science Notes 93/46, 1993.

[38] Kloks, T., $K_{1,3}$ *and* W_4-*free graphs*. Technische Universiteit Eindhoven, Computing Science Notes 94/25, 1994.

[39] Kloks, T., D. Kratsch and H. Müller, *Dominoes*. Technische Universiteit Eindhoven, Computing Science Notes 94/12, 1994.

[40] Kloks, T. and D. Kratsch, *Listing all minimal separators of a graph*. SIAM Journal on Computing **27** (1998), pp. 605–613.

[41] Kloks, T., D. Kratsch and H. Müller, *Finding and counting small induced subgraphs efficiently*. Information Processing Letters **74** (2000), pp. 115–121.

[42] Kloks, T., C. Lee, J. Liu and H. Müller, *On the recognition of general partition graphs*. Springer, LNCS 2880 (2003), pp. 273–283.

[43] Kloks, T. and Y. Wang, *Advances in Graph Algorithms*. Manuscript on viXra: 1409.0165, 2014.

[44] Kratsch, D., T. Kloks and H. Müller, *Measuring the vulnerability for classes of intersection graphs*. Discrete Applied Mathematics **77** (1997), pp. 259–270.

[45] Kruskal, J., *Well-quasi-ordering, the tree theorem, and Vazsonyi's conjecture*. Transactions of the American Mathematical Society **95** (1960), pp. 210–225.

[46] Krylou, Y. and A. Tchernov, *On the problem of domino recognition*. Electronic Notes in Mathematics **24** (2006), pp. 251–258.

[47] Dom Van der Laan, H., *Het plastische getal, XV Lessen over de grondslagen van de architectonische ordonnantie*. Leiden, E. J. Brill, 1967.

[48] Lekkerkerker, C. and J. Boland, *Representations of a finite graph by a set of intervals on the real line*. Fund. Math. **51** (1962), pp. 45–64.

[49] Micali, S. and V. Vazirani, *An $O(V^{1/2}E)$ algorithm for finding maximum matching in general graphs*. 21th Annual Symposium on Foundations of Computer Science (1980), IEEE Computer Society Press, New York, pp. 17–27.

[50] Minsky, M., *Computation, finite and infinite machines*. Prentice Hall, 1967.

[51] Minty, G., *On maximal independent sets of vertices in claw-free graphs*. Journal of Combinatorial Theory, Series B (1980), pp. 284–304.

[52] Moon, J. and L. Moser, *On cliques in graphs*. Israel Journal of Mathematics **3** (1965), pp. 23–28.

[53] Moser, R. and G. Tardos, *A constructive proof of the general Lovász local lemma*. Manuscript on ArXiv: 0903.0544, 2009.

[54] Nash-Williams, C., *On well-quasi-ordering infinite trees*. Proc. Cambr. Philos. Soc. **61** (1965), pp. 697–720.

[55] Nienhuys, J., *De Bruijn's combinatorics*. Manuscript on ViXra 1208.0223, 2012.

[56] Opatrný, J., *Total ordering problem*. SIAM Journal on Computing **8** (1979), pp. 111–114.

[57] Rem, M., *Algorithmen*. (eds. Van Oers and J. Peters) Technische Hogeschool Eindhoven, College Dictaat 2.292, 1981.

[58] Seidel, J. J., *Geometry and Combinatorics*. (Corneil and Mathon eds.) Selected works of J. J. Seidel, Academic Press, 1991.

[59] Shannon, C., *A theorem on coloring the lines of a network*. Studies in Applies Mathematics **28**, Wiley, (1949), pp. 148–152.

[60] Shi, L. and S. Ólafsson, *Nested Partitions Method, Theory and Applications*. International Series in Operations Research & Management Science, Vol. 109, Springer, 2009.

[61] Shoemaker, A. and S. Vare, *Edmonds' blossom algorithm* . Technical Report Stanford University CME 323, 2016.

[62] Spencer, J., *Asymptotic lower bounds for Ramsey functions*. Discrete Mathematics **20** (1977), pp. 69–76.

[63] Sumner, D., *Graphs with 1-factors*. Proceedings of the American Mathematical Society **42** (1974), pp. 8–12.

[64] Szemerédi, E., *Regular partitions of graphs*. Technical Report, Stanford University STAN-CS-75-489, 1975.

[65] Tarjan, R., *Decomposition by clique separators*. Discrete Mathematics **55** (1985), pp. 221–232.

[66] Thomassen, C., *Every 4-uniform 4-regular hypergraph is 2-colorable*. J. American Math. Soc. **5** (1992), pp. 217–229.

[67] Tomita, E., A. Tanaka and H. Takahashi, *The worst-case time complexity for generating all maximal cliques and computational experiments*. Theoretical Computer Science **363** (2006), pp. 28–42.

[68] Vatter, V., *Maximal independent sets and separating covers*. American Mathematical Monthly **118** (2011), pp. 418–423.

[69] Williams, V., J. Wang, R. Williams and H. Yu, *Finding four-node subgraphs in triangle time*. Proceedings SODA'15, Proceedings of the 20th Annual ACM-SIAM Symposium on Discrete Algorithms (2015), pp. 1671–1680.

[70] Whitesides, S., *An algorithm for finding clique cut-sets*. Information Processing Letters **12** (1981), pp. 31 32.

[71] Wolk, E., *The comparability graph of a tree*. Proceedings of the American Mathematical Society **13** (1962), pp. 789–795.

[72] Wood, D., *On the number of maximal independent sets in a graph*. ArXiV 2011.

[73] Xiao, M., T. Kloks and S-H. Poon, *New parameterized algorithms for the edge dominating set problem*. Theoretical Computer Science **511** (2013), pp. 147–158.

[74] Xiao, M. and H. Nagamochi, *Exact algorithms for maximum independent set*. Information and Computation **255** (2017), pp. 126–146.

Index

(a, b)-dominating set, 46
5-wheel, 28
8-wheel, 33
$G \times H$, 23
$G + H$, 23
$G - W$, 5
$G - x$, 5
GF[2], 187
$G[W]$, 5
H-free graph, 14
$K_4 - e$, 18
K_n, 23
$K_{m,n}$, 12
MS_1-expressible, 127
MS_2-expressible, 127
$N^-[x]$, 256
P_3 - convex, 121
P_3 - cover, 95
W - hierarchy, 90
W_4, 104
$[n]$, 7
AT - free, 204, 282
$\bar{N}(A)$, closed neighborhood in dependency graph, 48
$\Lambda(S)$, 72
NP, 39
NP-complete, 39
$\tilde{X}$, 170
$\alpha(G)$, 13
χ - bounded, 285
χ-bounded, 192
$\chi(G)$, 13

$\chi'(G)$, 40
$\chi_r(G)$, 81
$\delta(X)$, 150
$\delta(X, Y)$, 164
ϵ-regular, 54
ϵ-regular partition, 55
$\gamma(G)$, 317
$\gamma(A)$, 317
$\gamma^i(G)$, 317
$\gamma_{a,b}$, 46
∞, 236
$\mathbb{Z}^{\geq 0}$, nonnegative integers, 151
$\mathbb{E}$, expectation, 54
$\mathbb{N}$, the set of natural numbers, 2
$\mathcal{C}$ - governed decomposition, 197
$\mathcal{F}$ series parallel tree, 267
$\mathcal{F}$-free graph, 97
BFS-order, reversed, 51
DH-width, 227
cc(G), 175
index(π), index of partition, 56
rank(X), 187
$|\pi|$, size of a partition, 58
MOP, 321
μ, measure of carvings, 164
$\nu(G)$, 96
$\omega(G)$, 15
$\oplus$, 84
$\otimes$, 84
$\phi(G)$, 72
$\sigma(G)$, 75
$\tau(G)$, 218

θ_e, 319
tw(G), 134
ϵ-dense, 257
a, b-separator, 10
b(G), 135
d(X, Y), 54
d(x), 3
e(X, Y), 54
$e = 2.718 \cdots$, the basis of the natural logarithm, 43
k-alternating path, 261
k-degenerate, 29
k-uniform hypergraph, 34
$k \cdot K_n$, 23
p-carving width, 151
p-distance, 154
p-length, 152
q(G), 39
s(G), 73
t - embeddable, 244
x, y - path, 5
1-1 map, bijection, 150
2-coloring problem for hypergraphs, 34
2-connected, 166
2-player game, 112
2-tree, 263
3-partition problem, 310

acyclic, 264
adjacent, 1
alignment, 121

anti-twin, 86
antichain, 305
antihole, 191
antimatroid, 290
antipodality, 151
antipodality, of p-range $\geqslant$ k, 152
antisymmetric, 30
arity, 127
asteroidal set, 282
asteroidal triple, 282
atomic formulas, 127
automorphism, 123

backward edge, 244
bad event, 42, 43
bad sequence, 214
bag, 255
bandwidth, 205
Bell number, 139
Berge graph, 191
Bertrand's postulate, 321
betweenness, 290
betweenness constraints, 31
bias, 158
biconnected, 270, 276
binary relation, 30
binary tree, 229
bipartite graph, 12
blossom, 98
bond, 160
bond-carving, 164
Boolean function, 296
bottleneck, 20
bottleneck domination problem,
    20
bramble, 135
bramble number, 135
branch, 163
branching, 243
bridging operation, 125
Bron-Kerbosch algorithm, 22
Brouwer, L. E. J., 213
bull, 103

cactus, 145
Cartesian product, 30, 317
carving, 150
carving width, 150
castling, 113
Catalan number, 134
categorical product, 312
caterpillar, 287
Cauchy-Schwartz inequality, 57
cell - completion, 264
chain, 97
chain graph, 60
chomp, 122
chord of a path, 5
chordal embedding, 134
chordal graph, 129, 317
chordless cycle, 129
chordless path, 188
chromatic index, 40
chromatic number, 13
circle graph, 145
circle, in the Euclidean plane,
    145
circuit, 98, 159
claw, 14
clique, 14
clique - width, 221
clique separator, 74
clique tree, 132
closed neighborhood, 3
closed under complementation,
    84
closed walk, 151
cluster chromatic number, 205
cluster coloring, 205
cluster number, 205
co-NP-complete, 60
cobipartite graph, 134
cocomparability graph, 82, 283
cocomponent, 310
codeword, of type k, 251
cograph, 84
color classes, 12
coloring, 13

comparability graph, 82, 248
complement of a graph, 7
complete bipartite, 12
complete graph, 221
complete multi digraph, 255
complexity, 38
component, 7
composition, of functions, 197
computable function, 89
conference matrix, 299
conflict - free coloring, 195
confluent, 262
connected, 4
connected P_3 - game, 122
connected partition, 207
connective symbol, 126
consecutive clique arrangement,
    111, 176
consecutive ones property, 111
constant, 89
contraction, 79
convex geometry, 290
cotree, 85, 230
countable, 2
cover relation, 291
cross-free set-system, 150
crossing separators, 146
crossing subsets, 150
cut, 107
cut matrix, 187
cutsize function, 251
cutvertex, 10
cutwidth, 247
cycle, 6

DAG, 31
dangling, 11
data compression, 15
decision problem, 38
decision tree, 296
decomposition tree, 230
defective chromatic number, 212
defective coloring, 212
degeneracy, 29

degree, 3
density, 54
dependency graph, 44
depth, of Boolean function, 297
diamond, 18, 104
diamond-free, 18
digraph, 31, 246, 255
direct product, 312
disc, 159
disconnected, 4
distance hereditary, 125
distance hereditary-graph, 188
distance in graphs, 5
diversity, 197
dominating pair, 82, 204, 290
dominating set, 20, 82
domination, 317
domination in digraph, 256
domination number, 296
domino, 104, 189
dual, of a plane graph, 151

edge, 1
edge - lift, 247
edge clique - cover, 175, 237, 319
edge clique cover - number, 319
edge coloring, 40
edge domination, 89
edge intersection - graph, 174
edge intersection - model, 174
edge-contraction, 103
efficient set, 157
embedding, 321
empty graph, 1
endpoints of a path, 5
epithet: Meriam Webster's word
    of the day on 12 June 2018,
    80
equality symbol, 126
equator, 159
equitable partition, 55
equivalence cover, 39
equivalence graph, 39
equivalence relation, 2, 105

Erdős - Pósa property, 270
even separating triangle, 107
event, 43
exceptional class, 55
experiment, 43
expression, 126
extremal graph theory, 54

face, 321
factorizing order, 181
false twin, 85, 231
Fano plane, 46, 62
Farkas' lemma, 240
fast matrix multiplication, 60
feedback vertex set, 89
feet, 287
finite, 2
fixed parameter algorithm, 89
flip, 123
flower, 98
folklore, 84
forbidden induced subgraph, 97
forest, 11
forward Ramsey, 199
Fourier transform, 301

gap embedding, 216
gem, 105, 189
generalized decomposition, 197
Glenn Miller, 275
gluing operation, 125
graph, 1
graph algebra, 125
graph isomorphism, 2
grid minor theorem, 271
groove, 163

Hadamard matrix, 299
Hallian graph, 73
Hamiltonian cycle, 127
Helly property, 176, 270
hereditary property, 132
Higman's Lemma, 214
hitting set, 135

hole, 189
homeomorphism, 260
homomorphism, 305
house, 7, 189
hyperedge, 34
hypergraph, 34
hypergraph, k-uniform, 42
hypergraph, 2-coloring, 34
hypergraph, 2-coloring problem,
    42
hypergraph, regular, 46

immersion, 246, 259
incidence, 1
independence domination, 317
independence ratio, 314
independent domination, 255
independent set, 12
index, of a partition, 56
induced subgraph, 4
injective function, 215
injective map, 215
inside degree, 107
interlacing, 298
intersection graph, 133, 145, 319
interval, 283
interval graph, 111, 133, 175
intolerable clique, 238
invariant, 26
isolated vertex, 24, 217

Johan Cruijff, 112
join, 84
joke — a kind of, 174

Kőnig's infinity lemma, 213
kernel, of a flip, 123
Kronecker product, 312
Kruskalian decomposition, 197

labelling, 215
ladder, 122
laminar set-system, 150
layers, 322

layout, 247
legal move, 112
length of a path, 5
limb, 168
line, 320
line segment, 145
linear order, 31
linear time, 33
linegraph, 14
linked layout, 249
log, 49
logical symbol, 126
loop, 8
Lovász Local Lemma, 42
Lovász' local lemma, 44

Möbius inversion, 300
matching, 96
maximal clique, 22
maximum clique, 22
maximum flow, 74
maximum matching, 96
middle block, 267
minimal separator, 10
minor, 234
modular decomposition tree, 179
modular substitution, 203
module, 177
monadic second-order logic, 126, 127
monochromatic, 38
multi digraph, 256
multigraph, 262
multiset, 255
mutual independence, 43

Nash-Williams, Crispin St. John Alva, 214
natural numbers, 2
neighbor, 1
neighborhood, 3
neighborhood class, 197
net, 22, 293
nice tree - decomposition, 138

NIM - game, 114
nonadjacent, 1
nonedge, 41
NP-completeness, 35
null graph, 1

obstruction set, 235
odd antihole, 191
odd hole, 191
odd triple, 85, 102
one - way series - parallel triple, 265
open set, 151
oracle, 39
order, of bramble, 135
orientation, 34, 240
outerface, 321
outermost separating triangle, 109
overlap, 178

packing, 212
paired domination, 293
paper, scissors and stone game, 240
parallel composition, 262, 265
parallel extension, 265
parallel separators, 146
parametrization, of a class of graphs, 230
parametrize vs parameterize, 230
partial function, 49
partial order, 30
partial two - tree, 264
partitive family, 178
passive, 158
path in a graph, 5
paw, 80
payoff matrix, 241
pedant, 11
pendant vertex, 11
perfect elimination order, 131
perfect graph, 190
perfect graph theorem, 191

perfect matching, 112
permutation graph, 81, 282
permutation pattern, 145
pizza, 188
planar graph, 236
plane graph, 107, 236
point, 320
polygon, 147
polynomially χ - bounded, 196
portrait, 267
poset, 30
postcondition, 77
predicate symbol, 127
principle of diminishing returns, 249
probability, 43
probability distribution, 240
probability space, 43
probe module, 223
projective plane, 320
proper coloring, 81, 192

quantification over subsets of edges, 127
quantifier, 126
quasi-order, 213

random generator, 51
random variable, 43
rank, 81
rank of hypergraph, 34
rankwidth, 149, 187, 232
recognition of graph classes, 105
red maximal independent set problem, 239
refinement, of a partition, 55
reflexive, 30
region, 151
regular, 3
regular black vertex, 100
regularity lemma, 55
regularity lemma, constructive edition, 61
representative of a graph, 105

resampling, 48
retract, 305
Riemann hypothesis, 321
Ringel's conjecture, 212
Robertson chain, 260
Robertson-Seymour   theorem,
    235
root, 215
rooted digraph, 266
round set, 153
routing tree, 150
run - out, 213

safe reduction, 92
scanline, 82, 146
screen, 137
Seidel matrix, 87
Seidel switch, 102
sensitivity, 297
sentence, 126
separating triangle, 107
separation, 270
separator, 10
series - parallel triple, 265
series composition, 262
series extension, 267
series parallel separation, 270
shallow decomposition, 197
Shannon capacity, 191
simple elimination order, 293
simple vertex, 293
simplicial vertex, 131
simply, 266, 279
sink, 32
skew - symmetric, 241
slope, 160
snake, 112
source, 32
spacelab, 177
spanning caterpillar, 288
spanning subgraph, 4
spanning tree, 4
special region, 109
sphere, 151

splendid decomposition, 197
split, 198
split graph, 217
split of a piece, 83
splitgraph, 228
Steiner tree, 139, 207
stickiness, 73
stringer, 271
strong immersion, 246
strong module, 178
strong perfect graph theorem,
    191
strongly chordal graph, 293
strongly polynomial, 151
strongly regular, 106
subcubic, 259
subcubic graph, 260
subdivision, ℓ times, 160
subgraph, 4
submodular function, 157, 249
subtree, 132
subtree property, 132
Sumner's conjecture, 212, 242
sun, 103
Szemerédi's regularity lemma, 54

tagging, 197
take - away game, 122
tensor capacity, 315
tensor product, 312
terminal, 140, 262
ternary tree, 150
thickness, 72
thickness function, 72
thread, 261
threshold - width, 218
threshold graph, 217
tile, 188
tilt, 159
toplogical sort, 31
topological minor, 259
total dominating set, 21
total order, 31
total ordering problem, 31

totally balanced, 292
touch, 135
tournament, 240
tournament game, 240
transitive, 30
transitive closure, 255
transitive orientation, 82
tree, 10
tree - decomposition, 138
tree - degree, 175
tree of cycles, 263, 264
tree of triangles, 263
tree-like structure, 129
treewidth, 134
triad, 164
triangle, 15, 106
triangle-free graph, 104
triangulation, 134
trivial module, 177
trivially perfect graph, 80
trivially perfect graphs, 80
true twin, 85, 231
truncation, 267
twin, 85, 189, 231
two - terminal graph, 262
two-graph, 85

uniform slope, 160
union, 84
universal set, 34
universal vertex, 77, 80, 217

vertex, 1
vertex cover, 89
vertex minor, 286
vertex ranking, 81
Vizing theorem, 40

walk, 151
weakest precondition, 77
weight of a triangulation, 147
weighted graph, 20
well - rooted tree, 244
well - simulated, 265

well behaved, 266
well-formed, 127
well-linked, 223
well-quasi order, 213

wheel, 96
width, of a connected partition, 210
wing, 100

winning strategy, 112
witness to irregularity, 62
witness tree, proper, 50

Eindhoven and Chengdu   Oct. 2021

Printed in the United States
by Baker & Taylor Publisher Services